Kriegsgefangener
War Prisoner

Joseph H. Reus
Major, USAF (Ret.)

Kriegsgefangener-War Prisoner- is a work of non-fiction. Persons, places, events and situations are from real life during the World War II era and are drawn from memory, since prisoners were forbidden to keep diaries, which if found during searches by German Security personnel might aid our keepers. Modern day research by German historians has enhanced the author's knowledge.

This book has been typeset in Book Antiqua

Book design by Atlantis Productions, Edgewater, Florida 32132

Library of Congress Number

Dedication

This book is dedicated to all those valiant flyers who gave so much to the countries they loved and served - on both sides of our combat. They were among the finest young men of their many nations. They are not friends and foes - after our wounds have healed - but honored Brothers in Arms. Though many have now died, they will always live on in spirit - from the beginning of the aviation era until nations no longer compete in air battle.

This is one flyer's attempt to open a tiny window on the actual day -- to -- day life and outlook of a prisoner in one German Air Force prison camp during World War II.

LIST OF ILLUSTRATIONS

TABLE OF CONTENTS

Acknowledgements

German historians, Horst Jeckel and Robert Keller, have provided information about and photographs of the BF-110 aircraft (commonly, but erroneously, called Messerschmitt or ME-110) and the 21 cm missiles they lobbed into American bomber formations. Both have made wide and continuing studies of the World War II airwar in southwest Germany with all of its many historical byways including the fates of the crewmen of the many bombers that were shot down in battle. Horst Jeckel's research has found names and provided pictures of Focke-Wulf FW-190 fighter pilots lost over the English Channel on 30 December 1943 when we were attacked and had to ditch. Their help has made this book possible. I owe them thanks and have enjoyed visits with them.

Robert Keller, also, is the author of the German language book, *Luftkrieg über dem Vogelsberg* or *Air War over the Vogelsberg,* (The area northeast of Frankfurt). Their search of the records at the Lauterbach police department uncovered the note I wrote when captured and led them to me in the U.S. This same note still exists in the police files at Lauterbach.

Ms. Helga Radau of Barth, Germany, the historian of Stalag Luft 1, collected the camp's records when the POWs left, as a young girl of twelve. She has the greatest knowledge of that place and time - and is an active historian about the medieval aspects of Barth and northern Germany when it was part of Sweden.

Dr. Martin Albrecht, Director of the Berlin Archives, has graciously given me permission to use pictures and data from his book about the Heinkel HE-162 *Volksjaeger* (People's Fighter), that was being assembled at the airfield (or Fliegerhorst Barth) near the camp. Since the aircraft did not have time in the waning days of the war to get into battle with the Allied bombers, we'll never know what their influence would have been.

I also wish to thanks my son, Curt G. Reus, US Navy (Retired) for his continuing help and encouragement - and the large group of people who, when hearing my many WWII anecdotes pressed me to put them all into print.

The excellent photography work has been done by Peter Jolly.

Joseph Reus- June 1945

Joseph Reus-2004

Color Plate 1

UNITED STATES ARMY AIR FORCES

8TH AIR FORCE

SECOND AIR DIVISION

SECOND COMBAT WING

445TH BOMBER GROUP (H)

703RD BOMBER SQUADRON

COMMANDER - MAJOR JAMES M. STEWART

Goldfish Badge- Chapter 5

(Top)
US Army Air Force Navigator's Wings WWII
(Bottom)
and POW Camp made wings
Chapter

1
Combat Crew Training

After graduating from various advanced flying training schools across the United States from Florida to California our newly-minted, gold-plated second lieutenants for different crew positions were sent to Gowen Field at Boise, Idaho, to form bomber crews with the non-commissioned officers (NCO's) who would be our gunners, flight engineers and radio operators. In our case it was to create the B-24 aircrews of the 445th Bomber Group. The pilots, and some of the co-pilots, had been through flying training in the Liberator aircraft at a previous transition airbase specializing in this plane's handling - many at Smyrna, Tennessee.

The navigators and bombardiers on arrival were given turret gunnery courses so they could effectively operate the defensive weapon systems on the nose or other turrets of the later model Liberators. For our coming crew training, though, we were to use B-24D's which had no nose turret, but had a plexiglass paneled nose. The "D" model had only three flexible, or socket-mounted, manually operated, .50 caliber machine guns in the nose. It did, however, have the very effective and battle-hardened Martin upper turret and the rugged Consolidated hydraulic rear turret, plus the retractable ball-turret - each with twin fifties. The single side guns, or waist guns, were fired from an open window on either side of the plane. Later in combat flying we found that the ball-turret guns were rarely if ever used. The ball-turret also had to be in the upper, retractable, position for landing and take-off, because it would not otherwise clear the ground. Later, a gun mount in some Liberators was added at the front end of the camera hatch that was located in the floor behind the waist-gun positions. The hatch could be opened and the manually operated gun could be fired downward at targets not reachable by the tail turret.

The Consolidated hydraulic rear turrets survived so well in actual combat, even after receiving heavy gunfire damage, that they were later added as nose turrets in B-24J's and later aircraft to replace the much more damage-prone Emerson electrical turret. My later experience in this same turret was borne out vividly when it was shot to pieces around me by a Focke-Wulf FW-190. Most of this gunnery training, on the ground of course at Gowen Field, was learning to fire from a turret mounted on the back of a Dodge truck chassis. Lots of fun, but very noisy. The range was well away from other school buildings, naturally.

The moving targets used were made from rectangular cloth sails mounted between two metal posts fastened above un-manned jeeps. The jeeps followed an oval racetrack, between rails, and were protected from the heavy-caliber gunfire by earthen berms. This gave us a chance to shoot at the moving sail that retreated and advanced and went from left to right - then right to left at different ranges, or distances. The *lead*, or distance needed to fire ahead, to hit the moving target wasn't as changeable or as fast moving, as we later saw with enemy aircraft, but it was a good beginning and got the idea across. Firing a pair of real, heavy caliber machine guns with their very loud roar and recoil gave us a tiny idea of what was in our near future. It was like a very large pin-ball game on steroids.

This is also the reason that airmen were always required to practice *skee*t shooting with shotguns. We were required to fire a minimum of one hundred rounds per year throughout our service - to keep this lead in mind. We also, while at Gowen Field, flew several short-duration orientation flights with our new crews in B-24's.

The crews were then sent for 1st phase combat-crew training to a small airfield at Pocatello, Idaho, where we flew much longer, simulated combat, training missions to learn to fly as a crew.

Back then we were out in the middle of nowhere with no big city lights for distraction from our main goal of rapidly learning to be professional flyers, even though we were still very young and generally inexperienced. Some of the young gunners were recent high school graduates. The living conditions at Pocatello in tar-papered shacks were actually worse than I later had as a Pow - though the food was slightly better and in larger quantities.

The last and final, or advanced, training phase, for our squadron, was held at the municipal airport at Sioux City, Iowa. The airfields then were too small for the training of four heavy bomber squadrons together. Our 703rd Squadron, commanded by a thirty-five year old, Captain James Maitland Stewart, plus the 445th Bomber Group (H) headquarters, commanded by thirty-three year old Colonel Robert Terrell, were fortunate enough to be stationed there. Our other three squadrons were at various tiny airfields in very small towns from Mitchell, South Dakota, to Grand Island, Nebraska, but those towns were also an improvement over our *digs* at Pocatello.

While at Sioux City we flew many navigation training flights during the day and night, dropped many practice bombs, practiced formation flying, did gunnery training in flight, and had more ground training courses than I can remember. Our arrival in August, followed by a short pre-embarkation leave, quickly disappeared into November and heralded our departure.

A flight that sticks out in my mind, is a flight that featured a high-level simulated bombing run against a coastal area on the east side of Lake Michigan, near South Haven. The bomb run at altitude was scored by aerial photo.

Following immediately afterwards was a letdown so we could practice a *masthead* high low-level simulated attack against the tiny little aircraft carrier that operated near the Great Lakes Naval Training Station on the west side of Lake Michigan near Chicago. We came roaring across the lake at deck level, completed our practice bomb run, then pulled up over the ship's mast-head.

Then, instead of going on our way as planned, the pilot rolled into a right turn, dropped the landing gear and wing flaps, changed propeller pitch and appeared to be lining up for a landing with his big plane on the tiny aircraft carrier deck. We were far too big, of course, and our right wing would surely have hit their superstructure, or *island.* I believe the little carrier was built by decking over a small lake steamer. It was only used for practice carrier landings by the very small, Stearman-like, fabric-covered Naval biplanes we called the *Yellow Peril*. The Landing Officer on the carrier jumped off the deck into his net as we angled down and then turned away to the left at the very last moment. I guess the Navy was too embarrassed to report us.

One evening when we were scheduled for a night navigation mission, the flight had to be cancelled, because the crew using the plane ahead of us crashed on the final approach to the airfield as we watched and exploded just short of the runway, killing the crew. Instead of a flight, we had a *wake* at the bar in the Aviation Club of the Warrior Hotel in Sioux City. The hostess of this club was the girl friend of the pilot flying the plane that crashed.

Training was fast, but effective. Finally, after what seemed to be only a few weeks, we were ready for our *check-ride* as a bomber group. On the scheduled day, with inspectors from the Pentagon's Army Air Forces Headquarters, we took off, formed into a squadron formation and flew off to rendezvous with our other three squadrons. Everything went very well. We passed the exam and soon got our overseas orders.

Before we had time to sneeze, our ground crews (such as crew chiefs and their assistants, etc), were shipped by train to Fort Dix for the long sea voyage to England on the Queen Mary. There was a lengthy lead time needed to get them across the ocean to England and to our new base there. Now we were not only the flight crew - we were also the ground crew. I can remember climbing up on top of the engine cowlings helping to change the thirty-six big aircraft spark plugs on each of four twin-row, eighteen cylinder, Pratt & Whitney #1830 radial engines. You can also imagine what we looked like and smelled like when we washed the planes down with gasoline. We were too young to know it, but this was probably more dangerous that the war. We had to leave the planes in first class condition for the new bomber squadron that would take over from us after we left Sioux City. There was no one else but us left to do these jobs, so under the guidance of our flight engineer we all pitched in.

In a few days we received orders to pack-up and take the train to Lincoln Army Airfield, Lincoln, Nebraska, where we'd get our brand-new B-24H's. After the short train ride down to Omaha, Nebraska, we encountered a four hour delay before catching the next train that would take us and our mountains of baggage westward to Lincoln. We spent the time very sensibly nearby in a very large restaurant and bar called the Colony. It was so crowded that we never did get dinner. The waitresses brought us all the beer we needed, plus mountains of crackers and butter. When it came time to finally order we were too stuffed to eat a meal. We did leave a very good tip, though.

The next day, after little sleep in our tar-papered barracks there, each crew was given its factory fresh plane and the mountains of equipment and reams of documentation that's included. Our job was to *wring out* the plane, find out any faults, and get them repaired as quickly as possible by the factory representatives and USAAF mechanics on hand. My specific jobs included checking out the aircraft instruments and the drift meter. Some had to be done in flight; others checks were ground based. We had no laboratory equipment, such as used today.

To check a compass, we taxied out to a ground *compass rose* and had a *tug* tow the plane around in fifteen degree movements. With the engines running, I used a master compass to take reciprocal compass readings while on the ground in front of the plane. The pilot inside recorded similar reading at each stop from his magnetic and main compass. From the accumulated data, and the compass variation obtained from the local aerial map, I drew a sine wave curve and computed the errors if any at each position. This corrected data I later put on cards used for that purpose in a holder located at each compass location.

For airspeed correction, we flew down a railroad track at 500 feet elevation at 15 knot intervals, both ways, on each speed to cancel out wind effect. Using my stopwatch I calculated each time and averaged the two runs. Then I made up correction cards for each repeater airspeed indicator. By flying down the railroad tracks we also corrected alignment of the drift meter. The drift meter was used in flight to compute wind speed and direction and to correct our heading for cross-wind drift. These were the state-of-the-art *iron compass* days which wouldn't work very well with today's very accurate computer navigation systems. Of course, all vacuum instruments had to be checked by instrument technicians with their specialized equipment.

My 21st birthday arrived just before leaving Lincoln. There were no celebrations in wartime. The closest thing to that would be the night before. A group of ten of us, or so, including Jim Stewart, were in a hotel bar in Lincoln drinking Cornhusker ale, when the bartender leaned over to me and asked, "How old are you, kid?" I answered, "I'm 20, but I'll be 21 at midnight." He came back with, "Bullshit!" I showed him my ID card, containing my birth date which showed him that I really would be 21 at midnight. He said, "I'll be damned. OK, Kid." We all drank free; no bill. It was my only birthday present. I wouldn't have wanted any better.

On a Saturday morning while all of the bomber group were scheduled to stand inspection, a friend and I were dressed in our *pinks and greens* and service caps and arrived about an hour early. Having heard that one of the hangars down the row held a British RAF Spitfire and a German Messerschmitt BF-109E, we decided to kill time by walking down to see if it was true. We entered the closed large hangar door by the small panel door at the side. Ahead to the left were the two planes we came to see. Each was in perfect condition - with cowlings opened so we could see their large twelve cylinder twin-row engines. Naturally we'd never seen either of them before, and we were somewhat awed by real warplanes about which we'd heard so much.

While we were walking around the planes and climbing up on their wings to look into the cockpits, a sizeable group, of what we gradually recognized as a French enlisted training group, probably mechanics, in Class A uniforms. They slid back one of the large hangar door panels and marched in. They lined up on the right side of the hangar for what seemed to be an inspection of their own.

As we left the second plane, the BF-109E, the French NCO wheeled around and saluted. Not to be out-done, I returned the salute. Before I could walk on, the sergeant gave an order loudly in French. His men opened out into inspection formation and he marched down the front row. On a lark, I followed him at a rapid pace. Up and down each rank of the hundred or so men we went. My lieutenant friend had quickly picked up the spirit and followed with a straight face - as if he were my adjutant.

When our inspection was completed, and we were back in front of his troop, the sergeant again saluted. I returned the salute, and said, loudly, "Dismiss your men." The sergeant gave another order and his men, did a closed-ranks formation, then wheeled about and marched out the same door by which they had entered. Pity I didn't have white gloves and a saber.

I would also have liked to see the puzzlement on the proper inspecting officer's face when he showed up and found no one in sight. While we walked out of the hangar and up to our own inspection formation we had a laugh at the prank and wondered who got chewed-out for missing this inspection.

We didn't have time for many more stunts, because we had to get back to our barracks and begin to pack all the many, many things that we had to prepare for loading aboard our new aircraft. It all got completed, of course, and two days later after an early breakfast we took-off to begin the flight to Morrison Field at West Palm Beach, Florida, our overseas departure point.

After our arrival we were soon told that our crew would be one of the ten crews that had been selected as the lead element of the 445th Bomb Group for departure to our overseas base, but we weren't told where it would be. In fact, we didn't learn about our final destination until we'd reached Marrakech in Morocco.

Combat Group Training- B-24D Liberator
Glass paneled nose of the "D" model we trained on at
Sioux City, Iowa

Credit- J. Reus Personal Collection

Jimmy Stewart- Former CO of the 703rd Squadron
Taking crew out to plane in jeep before a combat mission.

Short Snorter Roll

During World War II, each flying officer put togther a *Short Snorter* roll of paper currency. You began the roll with the date on an American one dollar bill on which you had many of your friends inscribe their names. You added a banknote bill or piece of low value currency from each country in which you landed during your travels to and from the war zones where our battles took place. The resulting roll was a monetary travel map, or travelogue, that was much more interesting and colorful than a dull dry service record.

The propeller driven aircraft of that time had relatively short range - and no air-refueling capability - so for those flying their own aircraft to England to join the United States Army's 8th Air Force's battle against Germany, had many stops in as many countries along the way. Even though our bomber group flew the B-24 Liberator aircraft, which had far greater range and speed than the B-17 Flying Fortress, which was given the shorter northern over-water routes, we had to make stopovers in Puerto Rico, British Guiana, Belem, and Natal in Brazil, Dakar in Senegal, West Africa and Marrakech in Morocco, before finally arriving in Land's End in England with four new bills added.

When you stopped in a bar for a drink or even a beer at any officer's club, anywhere in the world, you could be challenged with, "Where's your Short Snorter?" If you couldn't produce it, and the challenger had his roll, you had to buy a drink for everyone seated at the bar.

Original World War II *Short Snorter* Roll signature
Capt. James "Jimmy" Stewart
CO 703rd Squadron 445th Bomber Group

2
Overseas Travel

In the pre-dawn darkness of 10 November 1943 Lt Charles L. Torpey's crew, which I navigated, was the third aircraft in line for take-off. Each plane also carried four priority passengers, such as the four squadron commanders of the ten planes involved in the first wave, plus group and squadron staff officers for operations, aircraft maintenance, communications, intelligence, flight surgeon, etc. Many were not flying officers, but I'm sure all were very glad to avoid the crowded *Queen Mary*, which in wartime was not exactly a luxury cruise liner, but more like a very large cattle barge. Also, it was important that these staff officers get to our new base as soon as possible and get set-up for our coming operations. Our group of ten planes, as previously noted, was selected as the advanced air element of the 445th Bomb Group. I still remember the brilliant red sky that opened as we climbed into the sunlight - an auspicious beginning.

Our first overseas flight leg left West Palm Beach bound for Borinquen Field, later named Ramey Air Force Base. It was located on the extreme northwest coast of Puerto Rico. Navigation was simple and easy. On the way you only needed a simple map to find your way, because there are so many easily identified islands and above-water projecting reefs to point out the route.

After we flew onward several hours, individually, we saw a huge cumulo-nimbus cloud barrier across our route. Because we were carrying four passengers, who had no oxygen masks or oxygen outlets, we couldn't attempt to climb over the weather. I gave the pilot, Charley Torpey a northeasterly heading to fly parallel to the cloud range while we looked for a passage through the spectacular wall. None showed, so after about ten minutes, I gave the pilot a reverse course. Then to have a little "sport", we handed out binoculars to the passengers, and asked them to let me know if they spotted any islands, so I could find my way. They all tried to help, while the gunners concealed their grins.

On the new southwesterly route, a few miles after passing our original turning point, we spotted a long, clear, "canyon" through the cloud wall with open sky beyond. We entered this giant golden cathedral with its spectacular view. Soon afterwards we passed between the many beautifully colored thunderstorms and dog-legged back to our route to continue onward. We landed shortly afterwards at the airbase in Puerto Rico and spent two nights there.

The first night's highlight was suggested by the older, wiser, and far more sophisticated Captain Jim Stewart, our CO. He flew in on one of the other nine planes and was with Lt Lloyd Sharrard's crew in the aircraft *Tenovus*. Jim's suggestion was frozen rum daiquiris before dinner at the beautiful officer's club. On the north side it had an open patio that perched on a high cliff wall that fell sharply down to a tiny beach far below along the Atlantic Ocean. The golden drinks were like liquid sunshine. I'm sure we were all smiles, too, especially for us *junior birdmen.* Now we all felt that we were finally off and running, although we did spend a second night there.

The next flight leg was also a simple over-water flight, with reefs and small islands to help point out the way to the South American coast and then up the river to the Royal Air Force base at Bridgetown, British Guiana, now Guyana. The base was thirty miles up the Demerara River from the Caribbean Coast. As we taxied in behind the *follow me* jeep and parked our planes we were met by British trucks that drove us from our widely separated parking ramp areas on the airfield to Base Operations. This gave us our first view of driving *to the left*. It also was our first, *don't drink the water area*, when we had lunch at the mess. No ice in the glass, only around the outside. The beer was not only good but safe. We'd left Florida's Fall warmth at West Palm Beach for the very real tropics. The change in heat and humidity was very noticeable.

A night's sleep was followed by the route that went next to Belem, Brazil - located about ninety miles up the Amazon River from the Atlantic Ocean, situated much like New Orleans. Belem is on the east branch of the Amazon River across from the large island, Marajo. Back then it was the source for many thousand caiman crocodiles that furnished leather for a thriving shoe and boot-making industry. Men rode out on horses carrying long heavy poles that were used to slam down and break the croc's neck so it could be quickly skinned. No bullet holes in the hide. The carcasses were abandoned and rotted in large piles we were told. Probably no one told them that the meat was good to eat. When I went back on an Amazon search expedition year's later, in 1948, I was told that there weren't many animals left.

The next flight was miles and miles of almost uninterrupted jungle with few savannahs of open land, and only a few small rivers, but no towns to break the monotony. It was a sea of trees. When we landed at Val De Caen Field we saw Brazilian army airfield guards dressed in German uniforms with the old World War I model 1916 German steel helmets and with long Mauser 98 rifles. They would have looked more like soldiers, however, if they had also been wearing shoes. Most were dark-skinned and probably of Indian descent.

Prior to the second world war the Brazilian and several other South American armies were trained and outfitted by German advisors. We also saw American Army engineering troops using American heavy equipment enlarging the runways and taxiways to accommodate ever larger numbers of American bombers on route to the build-up of the 8th Air Force in England and the 15th Air Force in Italy.

At Belem I became aware of Jim Stewart's wry sense of humor. We were continually warned to shake our boots out in foreign areas to avoid scorpions. I guess Jim noticed that I never remembered this warning. When I awakened the following morning and pulled on my boots, I got a painful shot in the big toe of my right foot. I yanked the boot off - and found a curled up pipe cleaner stuck in my toe. Jim had the bunk across from me in the open bay. Lesson learned. He pretended not to notice.

Our last stop in South America was Natal in northeast Brazil. It was our *jumping off* spot before the long over water flight across the South Atlantic. The day before, the pilot and I did our planning, such as the *go/no go* charts we used to check our fuel consumption to make sure we could make it to our destination, plus plotting the navigation charts and making up a navigation log. Here I began my first very long-range over-water flight by using the stars after the late-night take-off and later the Sun for celestial navigation. At first light, far out to sea I could see St Paul's Rocks directly ahead of the plane. The Rocks told me I was directly on course to our next landing at Rufisk Field at Dakar, Senegal, on Africa's Gold Coast. It also gave the pilot and I a chance to check our fuel consumption. It was a *go;* we were in good shape.

Longer-range four-engine bombers, such as the B-24, flew straight across the Atlantic Ocean. When possible, B-17's flew the north Atlantic route, because of their much shorter fuel range. Smaller twin engine aircraft, such as the B-25 Mitchell and B-26 Marauder, landed at Ascension Island out in the middle of the Atlantic, and south of the course we would take, for additional fuel. The route was quite a bit longer and took two days, but safer for them.

In the middle of the afternoon, as we approached Dakar, but were still about an hour out by my calculation, the pilot asked for my ETA, or estimated time of arrival. When I gave him our arrival time, he said that couldn't be correct. "With my binoculars I can see a brown line on the horizon that must be the coast of Africa." I checked my figures, but found no errors. I looked, too, also with binoculars, and saw the same brown line. We droned on, but never seemed to get closer. Finally we saw the actual coast, just at our estimated time, and began to realize that the *false* coast was actually the remnants of a dust storm blowing outward from the Sahara. We'd also have seen the same brown line if we'd looked either to the left or right across the sea toward either horizon. We were *babes-in-the-woods*, but we learned as we went.

Back then, our propeller aircraft didn't have the range of today's jets that I flew later (or have in-flight re-fueling). We flew only one leg per day due to our much slower speed and far longer time.

All I actually remember of Dakar was the concrete like beds covered in dust and the dinner aboard a British Navy heavy cruiser in port. The British admiral sent a dinner invitation to our group. No, unlike the British naval officers, we couldn't dress for dinner. All we had were khaki's. I'm sure we were invited because all of the naval officers, including the admiral, wanted to meet Jim Stewart.

The roast beef was probably much better than we'd have gotten in the mess at the airfield, anyway, and we got to eat off silver. We also had British assistance, or guidance, by some of the *lads*, on a short *pub crawl* in Dakar. Instead of a very large taxi, we had one big GI truck and driver for the whole gang.

The next flight was from Dakar, Senegal, northwards to Tindouf, Spanish Morocco, or Rio D'Oro. It's just south of the Atlas Mountains across many miles of desert after the green belt that covers much of Africa's northwest coast. There, when the weather is clear, you can fly through a mountain pass into Marrakech. But the weather wasn't clear. We couldn't even see the snow on the mountain tops. For safety I had the plane turn westward along the mountain range toward Agadir on the Atlantic coast. Then I noticed that our British map, supplied at Dakar, was a reprint of a French map. Surprise. No one had thought to warn us that the elevations shown weren't in feet; they were in meters. Since each meter is more than three feet, I'm glad we hadn't tried to fly through the weather over the mountains and gotten a last big surprise.

At Agadir we turned north across the Atlas mountain foothills to a point which would allow us to turn eastward to Marrakech. Again with an under cast, I computed a let-down point and told the pilot he could begin a descent at 500 feet per minute, until we'd break out of the clouds and see the local airport. Luck was with me. The plane broke out of the clouds at 1500 feet with the runway visible about 5 miles ahead. The pilot became a firm believer in dead-reckoning navigation, based only on time, heading, and distance. For new young navigators, it's probably better to be lucky than smart. Of course we never explained how we did things. After we landed and checked in we made the it to the mess before it closed.

The crew spent three days at the Marrakech airfield, during which we were quartered in a former French farmhouse and saw captured former German *Afrika Korps* soldiers working as cooks and waiters at the mess and as laborers around the base. Since the tide of battle had already turned in Russia, I'm sure most of the German captives were glad to be with us instead of dying on the Eastern Front, even if they didn't earn the German *Winter Schlact Im Osten* medal for Winter Battle In The East.

On the next day there was time to spend an afternoon and evening sight-seeing in Marrakech. We visited the vast market square and its snake charmers with their cobras in wicker baskets. We saw the Arab barbers who rode around on a bicycle type vehicle. The Arabs customers sat on a stool on the street and were shaved without water or soap lather. Sharpening their razor was done on an ancient leather belt. Their sanitation was limited to wiping the straight razor on the lower part of the rather dirty-looking lower part of the barber's *jellaba.* or robe

The sultan was not at home in his palace, so we were permitted an escorted tour through the many public areas. We later had a drink in the bar and dinner at the Mamounia Hotel, where Jim Stewart, after the war, made a movie called, *The Man Who Knew Too Much.* This hotel was for field-grade officers, majors and lieutenant-colonels. The local "peasants", lieutenants and captains, stayed at the Mahgreb Hotel or the airbase. We were only transients and beneath the lot. So we stayed at the air base.

Finally, after several days, we got the order to take off very early in the morning for the long flight up the Atlantic Ocean around Spain and Portugal. The route was northward past the Bay of Biscay, and well offshore from the French coast. Along this route I did my second *for real* night celestial navigation by shooting the stars. It came out well, since we didn't have enemy aircraft contesting our passage. We couldn't have afforded the *toll*. The German Air Force had long-range Junkers JU-88 Nachtjaeger (night-fighter) twin-engine aircraft stationed along the west coast of France hunting for pigeons like us. In any case, all of our guns were manned for the trip with our young gunners wide awake.

No cannon-fire welcomed our passage. When daylight broke, and the faint light brightened, every bug-splat on the glass of the nose turret, where I then sat with my clipboard and map, seemed to be an enemy aircraft until I finally recognized it and released my breath.

Our route termination was to a point in the middle of the Irish Channel west of Land's End, the southwest corner of England. Our landing was to be at the RAF Station at New Quay. Being over water we descended safely through the cloud decks until at very low altitude we could finally see the water. We couldn't turn in over land, because the clouds were lower than the English cliffs at Land's End that bordered the sea. Since we had no radar or radio beacon, we had to fall back on *Darky*. This was an emergency guidance system used by the RAF for night or bad weather.

If you were lost, or had our weather problem, you called "Darky, Darky, Darky," on the Emergency Channel. Next, you flew a complete circle to the right, clockwise, followed by one to the left, counter-clockwise. This alerted the local British night-fighter ground radar station, or GCI. They gave us a radio call with a heading to fly, while sending out a Spitfire to join us. The Spit arrived and we followed it through a coastal valley and made a safe approach to our intended airfield. After landing through low clouds and rain, we climbed out, huddled under the wing in the frigid air while we stretched our legs, and were next taken by truck to a mess where we were fed. After eating and a short break, the pilot, co-pilot, and I received a weather and route briefing at Base Ops to our new home base at Tibenham in the county of Norfolk while our plane was being refueled. We were told that we would fly in a loose formation convoy to our area.

As I mentioned earlier, we weren't told where our new base was until we reached Marrakech. Our new base was about 70 miles north of London and near the North Sea. There were four planes in our convoy. The lead aircraft was a C-54 transport from our Air Transport Command that flew the route regularly. Each of us flew in sight off the wing of the leader. We stayed close because of the heavy clouds.

As we flew north of London in a cloud opening we saw an AT-6 Texan training aircraft collide with an ancient Gypsy Moth biplane. They must have been watching us, and didn't see each other in the bad weather. In any case, three 'chutes opened and disappeared into the clouds below as we passed. The gaggle we were in flew over each airfield in turn and the proper plane departed the group and broke off to spiral down to land. When we arrived at Tibenham, the weather was too bad to attempt a landing, so we landed a few miles away at Hethel, the home of the 389th Bomb Group.

After dinner at the mess and spending the night, I was awakened in the morning by a young RAF girl with a large mug of tea, British-style with milk and sugar. After breakfast the weather looked a little better, even if frigid. We still had only our khakis to wear. After checking with Ops we were given permission to take off for an *under the clouds* flight to our own base only a few miles away. We finally reached Tibenham on the morning of 23 November 1943.

Now our 703rd Squadron, led by Captain James M Stewart, under the command of the 445th Bomber Group (H), directed by Colonel Robert Terrell, was a member of the 2nd Combat Wing, of the 2nd Air Division, and of the 8th U S Army Air Force, and after thousands of miles traveled, we were the first 703rd aircraft to land at Tibenham.

As events turned out, each time I returned to Tibenham, it was always, *Hello and Goodbye*. This will be confirmed in later chapters.

Here was our timetable: 10 November from Morrison to Borinquen; 11th at Borinquen; 12th Borinquen to Georgetown; 13th Georgetown to Belem, Brazil; 14th Belem to Natal, Brazil; 15th at Natal; 16th Natal to Dakar, Senegal, Africa; 17th Dakar to Marrakech, Morocco; 18th , 19th and 20th at Marrakech; took off the night of the 21st and landed in England at New Quay on the 22nd; on the 22nd we flew on to land at Hethel, home of the 389th Bomb Group; and finally, on the 23rd we made the short jump to our new base at Tibenham. We were the first of a very long line of B-24's that landed at the tiny village at Tibenham.

German Fighter Aircraft, Focke-Wulf FW-190-A

Commonly known is Europe as the Butcher Bird. This model aircraft had four 20mm cannons, two 7.92mm machine guns (30 calibre). This one is fitted with two air-to-air or air-to-ground rockets.

3
Prologue To Capture

On 30 December 1943 I flew as navigator for Flight Officer William P Kane's crew on B-24H, #41-29123, named *Big Joe.* At that time Kane had as yet no assigned navigator, and my own crew, piloted by Lt Charles L Torpey, had not as yet been assigned to a combat mission. Captain James M (Jim) Stewart, commander of the 703rd Bomb Squadron of the 445th Bomber Group (H), asked if I would like to fill-in with Kane on his first trip. Since all I'd flown so far in England had been a few training flights, I was ready to "get my feet wet". As it turned out and as this chapter shows I got much more than my feet wet.

The mission was to Ludwigshafen, Germany, to bomb a chemical factory on the west bank of the Mainz River south of Frankfurt. I'm not sure now whether it was a synthetic rubber or gasoline plant, but I'm certain that it was then covered well by the intelligence briefing. It was called a Buna plant which extracted hydrocarbon chemicals from the poor German brown coal called lignite, the only coal within their borders. Germany's only other source of these chemicals and gasoline was the large combine of five oil refineries grouped at Ploesti in Romania. This complex, managed extremely well by a German air force general, kept Germany in the war, despite the large low-level bombing raid you've read about, but also through at least five more bombing raids flown by the planes of the 15th US Army Air Force from bases in Italy. Great damage followed by quick repairs must have been his specialty. Ploesti struggled on and on almost to the end of the war.

The United States had a very large supply of oil and a well developed refinery system. America not only supplied all of the oil and gasoline for its many industrial needs, supplied the needs of each of its armed forces in battle fronts around the World, and provided the fuels for all of the Allied Forces. It never had to resort to the gasification of coal from our large in-ground supply. The only real limitation was tanker transportation to where the fuel was needed. For much of the war German submarines sank many oil tankers in the Gulf of Mexico and in the Atlantic Ocean - and in return suffered unbelievable crew losses. Ninety percent of submariners never made it through the war. I've never heard of any naval force with such terrible losses. Our tanker losses in turn caused America to build fuel lines across much of the land from the oil fields in the southwest to the users in the East.

In England during the war all of us used much more bicycle power than gasoline to get around the bases. English civilians, like their German counterparts, also used many small cars and trucks that created their own *gas* by means of a *cooker*- type device that roasted wood chips to create fuel. It was a strange looking thing - like a small hot-water heater with fuel lines to the engine. They were rickety, but they evidently worked well enough - even if slowly.

The only *black* coal, such as our vast supplies of excellent anthracite, is mined from northeastern France. This Axis gasification process made both basic products that were so vitally needed in the German war industries and to power their large armed forces. If Germany had a better fuel source perhaps the war might have had a different outcome - or would have lasted far longer.

Seeing for the first time, all of the preliminaries, including mission briefing and the many other pre-flight duties, the whole process was as new and novel to me as the first day in flying school. What I'd seen previously had only been during training exercises. Even with some gaps I could now see how well the training mirrored actual combat duties. Now life had suddenly became much more real.

On this first mission before reaching Germany, while *spelling* the bombardier in the Emerson nose turret I almost fired at an American P-47 Thunderbolt fighter. I recognized it at the last second as it dove down in front of us over Holland. It was chasing a Focke-Wulf FW-190 Butcher-bird that went by too fast for me to get off a firing burst and I almost fired at the second plane instead of the first. Everything else had gone normally through the formation assembly, crossing the North Sea, the Zuider Zee, and covering the route to the German border. We flew on from turning point to turning point and through the actual bombing run uneventfully. I looked down and watched as our string of 500-pound bombs punctured large flaming gaps in the plant. Soon would come a rude awakening and a quick introduction to combat reality.

At *bombs away* only half of our bombs fell from the bomb bay. This meant that we had to *salvo* the rest of the bomb load by the pilot's electrical switch as I followed through with the manual salvo release in the nose near my station while in the turn over the river crossing. In our swing out we also probably scattered some of the *hang-up* bombs across the west side of the city of Mannheim that lay on the river bank's east side. Hope the people were in their shelters.

Before we could close the bomb-bay doors, the aircraft was staggered by one or more very close 88mm flak shell bursts. Number three engine trailed a long streamer of flame and smoke and shut down. Its fuel line was probably severed. The right vertical stabilizer also was shredded and heavily damaged, with virtually all of the cloth-covered rudder surface blown away. The right rear area of the plane was riddled with holes, but neither the tail-gunner, Staff Sergeant Ray Sleeper, nor the two waist gunners, Sergeant James Stillwell and Staff Sergeant James Bivins, was hit.

We continued the turn to the right to the group rally point, but our airspeed gradually fell off as we lost power. The pilot dropped the nose, but our speed was still too slow to keep up with the formation now above us. The bomber stream could not wait for stragglers. We were on our own. The engine fire gradually reduced to a trail of gray and black smoke after the engine was shut down and the propeller *feathered*. Control of the aircraft took the efforts of both pilot and co-pilot, due to the damaged rudder, but was possible. We'd also lost the use of some of our flight instruments such as the gyro compass.

I gave the pilot a quick compass heading for a direct route from our position back to the English Channel and England. He'd have to do the best he could with his iron compass, or *standby* magnetic compass, since I no longer had a reading on the gyro-compass indicator in the aircraft's nose at the navigation desk. The broken cloud deck beneath the plane prevented me from using landmarks to check our route, but probably also prevented the following enemy fighter aircraft from spotting us.

The clouds eventually thinned out somewhat over time and distance and we eventually began to see the Channel ahead. Landmarks showed that we were farther west than expected.

Kane, the pilot, asked the crew's enlisted men if they'd like to safely parachute into the French countryside before we got to the heavily defended coastal security zone where the men would be more easily spotted and captured. The French underground in other areas had helped many British and American airmen to evade capture and escape across the Pyrenees Mountains through Spain, or perhaps they could be picked-up by the special RAF Lysander rescue aircraft that landed on French meadows by day or night and returned to England directly.

Our chances of safely crossing the Channel were probably doubtful. The engineer/gunner, Sergeant Gil Lloyd, as spokesman and senior NCO asked the pilot if the officers were going to attempt the Channel crossing. When told that they were going to try it, he asked to go along for himself and the others. It was then agreed upon that all would stay with the plane, regardless of the condition of the plane and of our chances of safely reaching England.

We limped out across the sea, slowly losing altitude until a small formation of twin engine planes rapidly passed under us, going in a northwesterly direction. To our great relief, the planes were Allied Douglas A-20s, or Havoc's, returning from a low-level raid into France. I don't remember seeing if it was an American or Royal Air Force bunch.

The bombardier, Lt Robert Jones, had manned the nose turret back across Europe from Germany. Now he suddenly realized that he couldn't move his feet, that his electrically-heated flying suit, or *Blue Bunny*, as we called it, wasn't working. He'd been very cold, but now found that he was too numb to move his legs. He told the pilot of the problem. The flight engineer and I moved forward together and lifted him out through the small turret doors and eased him to the floor, then plugged him into my heater outlet, which was still working, to warm him and get him on his feet.

With a clip board and folded map I climbed into the turret so I could more easily spot the English cliffs I knew were ahead. Then as soon as I recognized a landmark we could turn toward the nearest airfield for an emergency landing. The pilot and co-pilot were on the radio, which also still worked, with the air-sea rescue organization of the Royal Air Force. Their base was about eighty miles west of us, near Portsmouth, but their crash boat was too far away to help. The weather was too poor and the seas were much too rough to attempt a landing with their small single-engine Walrus flying boats. Even taxiing through the large waves would endanger and probably sink the small pre-war rescue plane.

The weather under us was getting worse, with large low clouds rapidly scudding past, but I could occasionally see the rearing white-capped waves through the many holes in the clouds. As I scanned from side to side I spotted what looked like a fighter aircraft to the right and level. I called it out to the crew as the plane came around in a wide, sweeping turn toward us. I then recognized the large radial engine of the FW-190. We were just another straggler to be picked off.

There was no time to say anything else before the actual attack. I dropped the map board and checked that the turret electrical system switch was "on". When the aircraft ahead was rolling out of its turn and closing rapidly, I grabbed the handgrips and swung the gunsight across to center on what I thought was its engine, the biggest target, and fired the twin .50 caliber machine guns. I fired only one burst, but never had a chance to see what or even if I'd scored a hit.

The enemy machine guns and four 20mm cannons' fire raked the turret and blew glass, metal and plexiglas debris into a shower of pieces. The angled armor-glass panel inside of the Emerson turret in front of me held together, but became virtually opaque. There was a great deal of loud noise followed by the roaring of the wind through the gaps in the turret. I felt things hit me and ricochet all around me, but felt no pain. I didn't know that the enemy fighter had also been hit, or that he'd knocked-out our other engine on the right side, leaving us with no power on the that side. The waist gunners both called out that the fighter crashed into the sea below us. I couldn't actually see it go down from the my position through what was left of the turret. Now I believe that my gunfire had been higher than the engine I'd aimed at and probably smashed through the fighter's windshield. Since he took no evasive action I believed he must have been inexperienced. We weren't the "soft" target he expected; just lucky. He probably got more than he bargained for, poor lad.

Later, a German historian friend researched old Luftwaffe records and found the names of two FW-190 fighter pilots who had been lost over the Channel that afternoon while attacking lone American B-24 bombers. I have both their names and rank, but there's no way to find out which one attacked our aircraft. Both were new to the game. One was a Leutnant, or 2nd Lt, and the other was an Unteroffizier, which equated to our *buck* sergeant. Both were new arrivals with little experience.

The majority of German pilots were enlisted throughout all of WWII. They received the rank of Unteroffizier when they graduated from flight training. American pilots upon graduation from flying school, who were not commissioned as 2nd Lieutenants, were given the warrant officer rank of Flight Officer, as was William Kane, the pilot of this crew.

The roaring wind around and through the broken glass and metal of the turret kept me from hearing much of anything up front. Somewhere in this time I climbed out of the nose turret, found that I'd been hit in the leg by something, but didn't realize I'd been hit by a 7.92mm machine gun bullet in the lower right calf. I injected one of the quarter-grain morphine syrettes, we carried for this purpose, next to where the blood was seeping into my boot. I plugged in to the navigator station interphone to find out what was going on.

We were now in much greater trouble with control problems, with engine power only on the left side, and with the right rudder out of commission. We began losing altitude more rapidly. The pilot told us we'd probably have to ditch when we broke out beneath the clouds and could see the water. The bombardier and I crawled through the tunnel under the co-pilot seat and moved to the flight deck with the engineer and radio operator. I continued on to my ditching station back through the bomb bay catwalk to the waist gun area.

My ditching position was in the rear of the plane and my crew duty was to supervise the dumping of as much weight as possible and to get rid of anything that would probably break loose on impact and fly through the air as a dangerous missile. The first things to dump were the long belts of unfired ammunition from the waist guns and tail turret positions. The .50 caliber machine guns couldn't be removed from their mounts without tools, but the internal parts of the receivers and the gun barrels could be removed and were pitched out of the waist windows. Our heavy flak suits and helmets also went overboard with everything else we could spare.

We could not parachute into the turbulent sea, because with the crashing waves and the fog closing in to hide us, we wouldn't be found soon enough, or maybe not at all, but would die of exposure in the wintry Channel. The RAF called our plight, the *5 percent club*, because that's what percentage of their flyers was found alive after parachuting or ditching into the icy water of the English Channel or North Sea at this time of the year. We were later told that the exposure limit was only about an hour. Airmen back then had no survival equipment provided for exposure to the elements. Many died quickly in this way on both the Allied and German sides when they were shot down or had equipment failure.

We also weren't briefed or aware of the rescue floats in the Channel, and probably were too close to the English shore to spot one. These were floating waterproof capsules anchored to the bottom. They contained blankets and wraps to maintain bodily heat, plus water and food. Airmen could take refuge in them. Sometimes, I'm told, they contained German and Allied airmen at the same time. Rescue boats from England and France came out regularly, weather permitting, to check and empty them. You could wind up with the English or German rescuers - depending upon who showed up first. With the large waves they must also have given a very rough ride, but at least they'd be alive.

After we'd done as much as we could, we braced for the coming impact, although back then no actual provision, or even safety belts, were provided in the back of the aircraft, or even on the flight deck behind the pilot for ditching. Little if any thought or previous design work by either the Army Air Forces planners, the aircraft designers, or manufacturers, had gone into crew safety in this event.

The men in the back of a B-24 had to brace against the rear wall of the bomb bay, on either side of the door that led into it from the rear. With four men, there was hardly room for two on each side.

In our case, one of the waist gunners and the ball-turret gunner took this space, while I, with no other good options, decided to brace against the back side of the ball turret's metal retraction-rigging. The tail-turret gunner, Ray Sleeper, braced against me, between my knees. I "popped" one side of my *Mae West* and had Ray do the same. The other waist gunner was sent forward to the flight deck.

The pilot called to brace for impact while I glanced to my left and saw water racing by just below us past the small, rectangular window forward of and below the right waist window. It was the last thing I still see in my mind's eye. In the giant, irregular waves there was no way to pick a landing spot. The plane hit with a huge impact as it crashed head-on into a large wave. All I really remember is flying through the air and slamming into the water. The impact broke both my outstretched arms. Something I hit gouged much of the skin from my left cheek.

Ray Sleeper hit the water just beyond me. As I looked at him bobbing over and under the big waves all I could see was a bloody scalp. I thought he had a fractured skull. His hair had been ripped completely back from the forehead, but the scalp was still attached at the back of his head. I later found that his only other real injury was a broken collar bone. He appeared to be unconscious at first, but after a moment paddled with me the short distance to the floating aircraft wreck.

The aircraft nose had been completely ripped off in front of the pilots cabin, and the fuselage was snapped into a broken angle at the rear of the bomb bay. The big, husky bombardier, Lt Bob Jones, lifted Ray and me onto the top of the wing where he was standing, being careful to avoid the shredded metal where the wing flaps had been ripped loose. This is when I first realized that my arms were useless.

The two five-passenger bright yellow dinghies were stored under hatches in the top of the fuselage on either side above the wing. They were designed to pop-out and self-inflate on such an impact - or activate from a pulled T-handle inside of the plane, but neither dinghy emerged. The bombardier, with help from the pilot who'd climbed out of his left side window, by using the aircraft fire axe, managed to get the small dinghy hatches opened and then pulled the cords that inflated them. When they were launched from the right wing, Ray and I were helped into them. Fortunately the plane stayed afloat in the crashing waves long enough to do what had to be done. Although I didn't remember it, the bombardier later apologized and told me he'd handed me one of the small red paddles. He told me he'd cussed me when I let the paddle drift away in the waves. Neither of us knew that my arms were broken, and I that couldn't close my fingers.

Four of the men had been killed on impact and went to the bottom of the Channel with the sinking aircraft. They were: co-pilot Lt Chuck Clemmens, whose seat had been torn loose from its moorings and who had been crushed into the debris that was left from the instrument panel ahead of him when the entire front section of the plane was destroyed. The heavy armor-plate behind and on either side, of his seat, meant for his protection, had most likely contributed to the impact. Flight engineer/gunner Sergeant Gil Lloyd died on the flight deck behind the pilot; and ball-turret gunner Sergeant Bob Forsythe and waist gunner Sergeant Jim Stilwell were crushed against the rear wall of the bomb bay when the ball-turret retraction equipment tore loose and slammed into them.

Lt Bob Jones, the bombardier, was probably the least injured, and was probably the biggest factor in the survival of those who made it to safety. He even managed to fire the flares that were packed into the dinghies. Without them we probably would not have been spotted in the thickening fog and the darkening skies by the fishermen who came out and who pulled us aboard.

I have no idea of the time involved. I was "in and out". Someone lifted me up into a double-ended open fishing boat with a flat deck and covered me with fishing nets to break the wind and spray, though I did get frostbite in both feet. The quick medical treatment later saved my feet from further damage. By the time my other injuries healed in February my feet recovered.

The boat crew, after seeing that we were the only survivors after the aircraft sank, headed northward to land. On the stony, pebble beach where we landed, we were carried across from the beached boat to other men carrying a long house-painting ladder. It had boards on it and was used as one big stretcher to carry all six of us. Somehow we were carried up a path to the top of the chalk cliff and packed into a big British ambulance for the short ride eastward to the

B-24s © 2004
445th BG website
Used with permission
on page 12, there is a descriptive of the top photo.

Princess Alice Memorial Hospital in nearby Eastbourne. I never saw any of these men again or had a chance to thank them, as I would have liked to do. They probably risked their lives many times throughout the war to save airmen, German and Allied, from the cold sea, during all seasons of the year.

I remember laying on a cold, black, plastic-covered table while I had my flying suit and other clothes cut-away for a full-body x-ray to look for flak holes or other injuries. The table in the unheated room was colder than the sea. The first few days were a complete blank, but I gradually began to take in my surroundings and recognize the people in the beds around me in the surgical ward. All of the doctors and nurses were women except for the one very young doctor. The nurses aides were very young girls hardly out of school. The only male doctor was a recent medical school graduate quickly pressed into service on arrival, because all of the other male doctors had been sent to serve with the British armed forces around the world.

While in the hospital, I had the bullet removed from my leg, plaster casts put on both arms and right hand, skin-grafting from the stomach to the left cheek, and various other small patches. Even as a beginner the young doctor did very well, as my later results showed. Staff Sergeant Ray Sleeper and I were the most seriously injured, so we stayed at the hospital when the other four men, after a day or so, began the train trip back to our base at Tibenham. Ray had his scalp sewed back down and recovered without a mark showing.

His father, as a civilian employed by the US Navy on the island of Adak in the Aleutian Islands off Alaska, later showed me color movies of his recovered son. I was stationed there on a duty tour with a P-61 Black Widow night-fighter squadron at the time in 1947-48.

As our recovery progressed, and as we both were able to be moved we were taken by ambulance to the US Army's 38th Station Hospital at Winchester. This hospital's surgical ward was run in a much different manner than any other Service hospital I ever saw in later years. In the bed next to mine was a B-17 bombardier who had just received his fifth, or silver, Purple Heart medal award for combat injuries. He was given Scotch whiskey for an anesthetic, both mental and physical, and sat on a large rubber donut. He'd been hit in the behind by part of a propeller blade shot away from his number two engine by an enemy fighter while bending over the bombsight during the bomb run. He had two very large flesh wounds, but no internal injuries. He was in good spirits, but as nervous and jumpy as a pea on a hot stove, as you can well imagine - this being the fifth time.

The flight surgeon who ran the ward at Winchester furnished the other wounded airmen with English beer in large stoneware pitchers on their bedside tables with the same aim. The local beer didn't have to remain cold and was readily available. He said it was probably better than chemical anesthetics and was an excellent pain-killer. I had to have another patient pour me a glass, because I couldn't manage the large stoneware pitcher with my two heavy arm casts. While there I received another skin graft from the right hip to the lower right leg to cover the bullet hole that had been sewn closed earlier at the Princess Alice. From there, Ray Sleeper, when well enough, was shipped back to the US.

I was later taken to the US Army's 5th General Hospital at Salisbury, since 30 days was the time limit for patients at a station hospital. The larger 5th General was located in a taken-over former girl's finishing school on the highway south of Salisbury that led down to the resort town of Bournemouth on the Channel. My stay there, plus the time spent in the two previous hospital, stretched out to seven weeks.

During my bed-stay there I still had the duty of mail censorship even though the cast on my arms and right thumb kept me from signing my name, rank, and serial number to the envelopes as required for censors. I was paired with a tank captain, the former provost-marshal of Rabat, Morocco, who'd been brought up from Africa for surgery. He'd crank up my bed, put a pillow on my chest, and slit open the letters to be censored. He laid out the pages I read, as well as his own, then put the letters back in the proper envelopes and signed both. Glad he got that right!

You can imagine what the results would have been if he'd gotten them switched. It was a morning ritual, after the doctors came through with their morning visit. My eyeballs still functioned, and I guess we had to earn our pay.

Across from me was an old English army major with a broken hip, a "Colonel Blimp" type and an instructor at Sandhurst, the British version of West Point. We really enjoyed him. His dry humor really livened everyone's day and stay. I still remember most of his ribald English jokes and funny stunts.

We had few people to help us, so we were "buddied-up", to help each other. Later when I was allowed out of bed and went to the mess, to other wards, or labs, to the movie, or anywhere else, I traveled by wheelchair pushed by my partner, a Canadian Army lieutenant. I couldn't walk because of my wounded leg, and couldn't push the wheelchair because of the arm casts. The Canadian couldn't see, because both his eyes were still covered after surgery. We made a "terrible team" as we flew down the walkways, going and coming. I had to pay attention, because we traveled fast. It was, "Slow, right turn (or left), Now!" Needless to say, everyone learned to stay out of our way, because, *Watch Out, Here They Come Again*. Since the lieutenant couldn't see to shave, I did that job for him and he luckily retained both ears and nose. What a wild pair! We were both 21.

Newly promoted Major Jim Stewart drove the 150 miles down to visit me in his small, issued, British Consul car from Tibenham to bring my uniform to wear when released. The nurses (even the doctors) went gaga and gathered in groups when they saw Jim arrive and ask for me. We had dinner together, then he left for the trip back. He left me a letter officially requesting that I be re-assigned to his 703rd Squadron.

When I had the arm casts removed and was able to stand without the wheelchair, I was allowed to walk around the hospital, then walk into Salisbury, or to the country pub that was down the road toward the south coast. I even joined the nearby British Army artillery officers's mess for a fee of a few shillings so I could eat at their steak night and drink their beer. To go outside I was given an ill-fitting bright blue suit, a white shirt, and red tie. This was to identify me if I fell over so I could be seen as a patient and brought back to the hospital. Staggering was not mentioned, but was included.

After a final medical exam that included lifting a fire bucket full of sand, with one hand and then the other, and the question, "Does that hurt?" I met a medical board that told me I could go back to the States if I wanted. I told them I'd rather go back to my squadron and showed them the request from Major Stewart. Train travel was set-up for me to London and on to a replacement depot, called Washington Hall, at Chorley near Preston, northeast of Liverpool.

During the three days there, waiting for travel orders, I was asked by an older chaplain to assist him in a tour for an enlisted group to the nearby Standish House, Miles Standish's ancestral home and other scenic spots nearby. Miles was well known in our American history before the Revolution.

Before I left that base one very amusing event took place. The replacement depot base commander, an older Lt Colonel, drove in through the main gate one night, while coming home from the nearby pub while quite a bit under the weather - and then really under water. During the blackout he failed to turn left to follow the base road and drove straight into the *fire pond.* and out of sight. These ponds were created to provide water for fighting fires during air raids. I was told he came up, spluttering, "What happened, what happened."

After receiving my travel orders I took the train the next day back down to London, then north on the London-Midland railroad line to Tivetshall Station, south of Norwich. It was near the Tibenham base that housed our 445th Bomber Group (H). Before I left Washington Hall replacement depot I had sent a telegraphed message to Major Stewart, the CO, with the estimated time of my arrival.

Jim met the train and drove me back to our base. He said he was glad that I was okay and had come back to the squadron. Then in the conversation he asked what I wanted to do after completing the normal 25-mission combat tour. I told him I'd probably like to stay on. He then told me that if I'd sign on for another combat tour afterwards he'd pull me out after a few missions so I could become squadron navigator on his staff. We'd been friends since training days at Sioux City, Iowa, and I'd certainly have been happy to keep him as my leader.

On the way, I asked him if I could fly one more mission before taking the mandatory *rest leave*, in Scotland, that was required after a hospital stay as long as mine had been. I wanted to "get back on the horse as soon as possible". He said that it wasn't possible, unless I could get a one-time waiver from our local flight surgeon, Major Black. When we got to the base, Jim called Black and invited him to dinner at the Club without mentioning me. He agreed.

We met the flight surgeon at our mess and club located back in a tree-covered part of a farm area. You had to open and close a large farm gate and go around a barn to get to it. This rural route was for air raid protection. I was still 21 and the other two were in their thirties. I answered questions only. After a few drinks and dinner, Jim asked Major Black if he'd okay a one-time flight. He agreed and said he'd send the waiver to Jim's squadron headquarters to be put it into my file.

The next day I got a message over the Tannoy PA speaker at the Nissen hut I'd been assigned to. Jim asked me to come down to the squadron. I borrowed a bike from one of the hut mates who was away and pedaled down. When I arrived Jim told me that Lt Pete Abell's navigator was sick and that I could take that slot if I wanted to, and that it was agreeable to Pete, a friend from training days at Sioux City.

What I didn't know, since I'd so recently returned to base and was out of touch, was that I'd unknowingly jumped into, *Operation Big Week*. It was to be a six-day bombing campaign extending from 20-25 February 1944, aimed at the German and Axis aircraft industry all the way across Europe. It was later recognized as the greatest air battle in history. Although the 8th Air Force had its biggest losses of the war in this campaign, we contributed in large measure to the final victory in the air war. Although we did not slow aircraft production as had been hoped, this multi-day air battle so decimated the German day fighter force that they went into a downward spiral from which they never recovered. Historians also later picked *Unternehmen Zitadelle* (*Operation Citadel*), the enormous German-Russian ground battle at Kursk, as the greatest land battle ever, and the American-Japanese naval Battle Of Leyte Gulf as history's greatest sea battle.

Lt Kane was later credited with the first "successful ditching" of a B-24 aircraft. Yeah, we lost only half the crew, because of the aircraft design and safety equipment lack, but through no fault of his.

As a personal note, as the survivor of an aircraft ditching at sea, I was awarded the British Royal Air Force *Goldfish*. It's an unofficial cloth emblem on a small black patch with a small fish outlined in gold with a large white wing. It's still on my WWII uniform. They are quite rare - there were few survivors who came down at sea.

Goldfish Club Badge

Actual size -1 1/2″ x 2″

Please see Color Plate 2 for illustration

The Goldfish Club was founded in 1942 by Mr C.A. Robertson, Chief Draftsman of the company, P.B. Cow &Co. Ltd. It was one of the world's largest air-sea rescue equipment manufacturers.

The Club was formed as an exclusive club for airmen who had survived a wartime aircraft ditching and who owed their lives to the successful use of the "Mae West" inflatable life preserver, a rubber dinghy, or other type of life raft, etc.

The Goldfish Club Badge shows a white-winged goldfish flying over two symbolic blue waves. The name `goldfish" was selected with gold for the value of life and fish for the sea.

Each new member of the Goldfish Club was presented with a heat-sealed, water-proof membership card and an embroidered badge with the financial backing of P.B. Cow, Ltd.

Due to wartime regulations (WWII) production of wire badges was prohibited and all cloth in England was severely rationed. These problems were overcome with silk embroidery substituted for wire upon black cloth cut from old evening dress suits that were sent in by readers of the London Daily Express after an appeal by columnist William Hickey. Uniform dress regulations prohibited the wearing of the Goldfish Club Badge on British and American uniforms. The badge was generally worn by Naval aircrews upon their "Mae West's;" many RAF and USAAF aircrewmen placed their badge under the flap of their left hand uniform pocket.

My "Goldfish," without the two blue waves underneath the fish, was presented to me by two RAF officers as I lay in a hospital bed at the 5th General Hospital, Salisbury, England, recovering from injuries received in battle and in ditching a B-24-H aircraft in the English Channel while returning from a bombing mission at Ludwigshafen, Germany on 30 December 1943 five to ten miles off Beachy Head near Eastbourne.

I later learned from my Royal Air Force friends while I was a POW at Stalag Luft 1, that the Goldfish Club was also known as the "five-percent club," because it was estimated that only five-percent of those who came down in the English Channel or in the North Sea in which it emptied were rescued alive unless they were picked up within an hour - due to the ice-cold sea temperature in both Winter and Summer. Six of our crew out of the ten aboard were picked up alive within that hour by a fishing boat that saw our flares.

4
Early Call to Take-Off

A bright flashlight, out of the dark - a gruff, whispered voice, "Lieutenant, Lieutenant. Are you Lt Reus?" My grunt must have sounded like, "Ok," because the voice said, "I'm waking Lt Abell's crew. Be back to pick you up in thirty minutes." This is the way a combat mission began, at 3:30 A.M. Because four officers from two crews, eight men in all, normally lived in the same Nissen hut, a flashlight and a whispered call was used to awaken individuals, in case the members of the second, or other, crew were not flying that day.

In this instance I had just returned from the hospital system after a seven-week absence and was no longer assigned to the crew with whom I'd trained in the U S, nor was I living in the same hut with the original crew's officers. I was also not a member of Pete's Abell's crew, nor living with his bunch. The *pick-up* driver must have had a very up-to-date map to find me, because I'd only slept in that bunk, or in that Nissen hut for two nights. To run a bomber base must have taken all sorts of competent men such as this, and far more people than the flyers knew about or rarely saw.

I turned on my small shielded flashlight, put on a coat over my pajamas, grabbed my shaving kit and left our hut for the latrine just a few feet away. It was located in a small, sunken brick building with wash basins, shower stalls, and toilets surrounded outside by a brick walk and a head-high earthen revetment - it was also our air raid shelter against low-flying, strafing enemy planes.

After leaving the warm bunk with its pile of covers I did much shivering in the pre-dawn darkness of that February morning as I hurriedly shaved in the empty building. I returned to the hut and got into the newly issued, electrically-heated, "blue bunny" flying suit and the other flying gear, topped by an olive drab jacket. It had all been collected only the day before from several supply sources on my return to the 445th's airbase near Tibenham, including a navigator's briefcase filled with the tools of my trade. I was missing one vital piece of military equipment that I was not aware of until it was brought to my attention later that day in Germany by the local police chief. Its substitute is on a very small piece of paper still in the archives of the Lauterbach police files.

I awaited the heavy *6x6* truck that would pick me up and stay with the crew at each stop until takeoff on the mission. It had steps on the back so we could climb in with all our heavy flying gear.

When the truck returned it had the six enlisted men: The engineer-gunner, the radio operator, plus the two waist gunners, the ball-turret gunner, and the tail gunner from their different barracks area. We didn't know then that the tail-gunner, Lonnie Reeves would not live to see a second sunrise or where each of us would spend the remainder of World War II.

After climbing into the truck we moved on to the Nissen hut where I said hello to Pete Abell, the pilot as he climbed aboard with the bombardier and co-pilot. Pete was a friend from training days at Sioux City. I was replacing Pete's sick navigator for this one mission. You can't fly an un-pressurized high-altitude combat mission with a head cold or blocked nose.

Next stop was the combat mess. It was separated from the non-flying mess and divided by an NCO at a table, because combat flyers were given the "hospital" food ration, or 1½ times the normal ration. We also got any fresh food that was available, again, under the same rules. Usually to get fresh eggs, though, you had to have a girl friend who lived on a nearby farm. They fattened us up for the kill.

To keep non-flying crew people out, flyers had to wear a sewed-on royal blue patch under their wings when they wore the green or olive drab uniforms to keep others from simply pinning on wings for the meal. Of course, in this case we were wearing flying clothes to breakfast. Earlier, non-flying men (and officers) had pinned on wings, when they went to town, also, to pretend that they were combat flyers and compete for feminine companionship.

Our older group executive officer, previously a non-flyer, took an aerial-gunnery course, just so he could wear the blue patch under his newly minted gunner's wings. *C'est la guerre*. I don't know if he ever tried to barge into the combat-mess. I also wonder if he ever actually flew a combat mission. If all the *paddle feet* from headquarters actually flew combat missions with us, we'd possibly have won the war much sooner, but the loss rate probably would have been far greater.

After a quickly eaten breakfast of *square eggs* made up from dry-egg powder, we had bacon, sausage, flapjacks, etc, plus lots of coffee Then we straggled out to the truck for the short drive to the large briefing room at the Group Operations building. On the mission we only carried the large fruit juice cans, but no solid food was provided. The eggs that day, even if powdered, were the last I'd see for more than a year and a quarter.

In wartime there were many forms of *gallow's humor* among soldiers, sailors, and airmen - and I'm sure, among all those in every hazardous profession as a form of *escapism*, in a very dangerous time. Our air war losses before we were provided with close fighter escort certainly qualified.

Our battle losses of killed, captured and wounded were so high and so continuous that Congress began to talk, with Churchill's encouragement, about changing our daylight formation bombing to a night bomber stream program much like the British Royal Air Force's Bomber Command. Both Germany and England had earlier tried daylight bombing, but each had to revert to night attacks due to severe aircraft and crew losses. Both countries also had populations only half of the U S size.

Fortunately, we were allowed to continue in daylight formation bombing until our 8th Air Force was able to grow large enough and skillful enough to decimate Germany's daylight fighter defense. If not, it's very probable that the war would have taken far longer. Back then night bombing was inherently less accurate and Hitler's late-war weapons might have come into being and caused a stalemate. We also know that it would have caused vastly greater combat losses on both sides. The Royal Air Force's bomber and aircrew losses continued unabatedly until the end of the war, even though its bombing accuracy could not be greatly improved with the equipment available.

Our 2nd Air Division, the B-24 Liberator bomber force of the 8th U S Army Air Force, alone suffered more than 6,700 killed in battle, plus many thousands more wounded or captured. Similar losses were sustained by the 8th's other two air divisions - the 1st Air Division's B-17s and the 3rd Air Division's combined B-17 and B-24 fleet.

The 15th Air Force, operating from Italian bases, although with much poorer living conditions, had smaller overall bomber losses, except when they attacked targets in Austria and southern Germany. Additionally, they faced fewer German fighters and those were of lower fighting capability. Germany, as its air defense fighters dwindled, quite naturally kept its best fighter units for home defense.

As a side light, over the door of our own 703rd Squadron Ops building, which we passed on the way, was a sign, nicely printed on a board in old English script. It said, "We Don't Bomb Germany, We Shower Them With Our Burning Wreckage!" Another squadron had a sign which read, "Getting Killed Is What We Do For A Living." I don't know if the other two squadrons in our bomb group had a similar *whistling in the dark* sign. I never noticed.

As we entered the large briefing room we gave up our wallets and other ID, except dog tags, to the intelligence clerk to put into the safe for us until we returned - if we did. If we didn't return to pick up the wallets and such, they were given later to an appointed summary court officer with our other belongings. He'd return them to our families with all our other possessions if we were killed or missing and didn't make it back from the mission. The wallets were properly "sanitized" to make sure a picture of a local girl friend didn't get sent back to a grieving wife. I did this job, as a summary court officer, myself, during the Korean war, even while flying combat missions in B-29's.

We then filed in to sit by crews for the mission briefing. At the proper moment the curtain was pulled back from the very large map on which the day's mission was laid out in colored string. We could see the usual route across the North Sea to Holland, the large shallow-water Zuider Zee, eastern Holland, and finally the German border. Then came the turning point at Dummer Lake. It was used very often, primarily because it was easily recognized by the pilots and navigators.

From there, after two more turning points, we were shown the Initial Point, or IP, from which the bomb run would be started toward the planned target. It was Germany's largest multi-engine BF-110, 210, and 410 manufacturing plant and was located just east of the town of Gotha. It's proper name was the Gothaer Waggonfabrik AG, or Gotha Vehicle Manufacturing Corporation. After a few words from the Group Commander, Col Bob Terrill, about the importance of this target and our part in the current bombing campaign against German aviation, the group's staff officers gave their general briefings. Each covered his part of operations, intelligence with expected enemy resistance, and weather expected. When each of these officers had finished his part of the plan we were given a final word by the Group Commander, dismissed, and we moved to our different rooms for specialized and more detailed, briefings. There was a room for the pilots & co-pilots, bombardiers, navigators, radio operators, gunners, etc, to cover their needed professional information in a more detailed and complete manner and to answer any questions.

As a navigator, after listening to the Group navigator's remarks, I sat down to rapidly compute the compass headings for each leg of the flight, with estimates of times at each turn point, calculations and other entries needed for my flight log using the weather sheet with its wind speed and direction that I had been given.

The bombardiers got such things as atmospheric pressures predicted so they could compute the true bombing altitude for the target areas, usually for the primary and secondary targets. This is of vital importance for bombing accuracy. They were also given information about bomb loads and bomb fuses to plan their bombing and ballistics data.

Radio operators got the list of radio frequencies to be used - on a "flimsy" sheet that told them when certain radio beacons would be on or off the air and how their call-signs would change with time. Their sheet was a more specialized one, with much greater detail, than that prepared for the pilots and co-pilots.

These changing radio beacon schedules made it more difficult for enemy aircraft to use the radio beacons for navigation during their attacks on Britain.

The pilots and co-pilots also got their copies for radio *homing* use during the complicated climb-out and formation assembly when we used the *splasher* and *buncher* radio beacons, for in-flight emergencies, and for the later return. They were also given charts that were used to show them their place in their *combat box* in the massed, multi-level formations we'd fly.

The *on-off* radio beacons were also set-up to prevent German attempts at mis-directing returning aircraft. The German Kriegsmarine, or war navy, usually had submarines in the English Channel and North Sea that sent out counterfeit radio signals to misguide returning planes. The beacon identification letters in Morse-code were hand *keyed* by the sub's radio man. This process was called *meakoning the beacon*, and did result, so we were told, in the loss of crews returning to England alone, mostly by running out of fuel over open water such as the North Sea. This applied mostly to RAF planes at night and in bad weather. We were warned regularly.

There was much more, really, than I can cover here. I mention all these needed planning items to give an inkling of the hugely complex structure that made possible bombing raids of such magnitude with the very large number of aircraft needed in that era. The success of large aircraft formations is even more amazing when you consider the young age and relative inexperience of the flyers.

Afterwards a stop was made at the building where the *ready-room lockers* that held our flight gear were located. We picked up parachutes, Mae-Wests (life-preservers), flak suits, etc.

I already had all my navigation gear, but I had to borrow the squadron Op's officer's parachute the day before. He told me to be sure to bring it back. Due to his larger size, I later found out that when I had to *jump* over Germany, the harness was much too large. The chest strap that should have been across my chest flew up and a buckle chipped a piece from an upper front tooth. Fortunately I didn't fall out of the chute when it opened with a bang. That wasn't unheard of either. My previous *fitted* chute was at the bottom of the English Channel from the previous mission. The harnesses were fitted for us by the parachute shop, by sewing through the straps. I forgot to get a new fitted chute the day before, in my rush to gather all the things I needed for my return to flight. A navigator uses a lot more than just maps.

The trucks made their last trips, this time to the planes. It was so dark, with rain and light fog, the drivers had to know where each plane was located. They usually got it right. The drivers, after unloading the bomber crew, parked behind the *hard stand*, or concrete pad, on which the plane was parked. Since it was cold and miserable, they huddled in their cabs near the heaters to stay warm and dry until the planes took-off. If the mission was delayed or scrubbed, they were available to bring the crews back to Base Op's.

The crews unloaded the truck, climbed into the planes and dressed in all the pieces of clothing and gear they weren't already wearing. Each crew member began his mental or written checklist. The gunners assembled their machine guns. Each gunner had his own barrel, or multiple barrels for turret gunner and tail gunners, plus the internal parts of the machine gun's *receiver*. Gunners cleaned, maintained and stored their own guns after each flight.

We each checked our equipment, and the items we used in the plane, except for the radio operator. He stayed off the air, because the German radio listening and direction-finder stations across the Channel were standing by and listening to learn whatever they could. If the radio operator turned the radio on, the carrier wave would go out and alert the enemy, even if the operator transmitted nothing. You can well imagine what the listeners across the Channel would have learned from the hundreds or thousands of carrier-waves. Later, however, we'd have to communicate, but in a different manner.

At the scheduled time the tower fired a specifically colored flare that had the specified number of *balls* from the many types available in various colors and balls, to announce *Station Time*, when each man must be at his aircraft station. Anticipation bred a quiet excitement. Next came the flare with colors for, *Start Engines*, *Taxi*, and finally, *Take Off*, each on a precise schedule and with a different flare color and pattern.

The flare pattern changed for every mission and for every airbase. Double-ball-red never changed, however, it was only used on return. It was the signal to the waiting ambulances that there was *Wounded Aboard*. The nearest ambulances then raced forward to get to the wounded men as soon as the plane stopped rolling.

The pilot, co-pilot, and navigator each had a copy of the *flimsy* sheet with the flare colors and timing. They backed up each other on the flares as they did with all timing requirements. Beforehand the pilots had used their formation chart to find the hard stand location of the plane they were to fly and next for the aircraft they were to follow through the taxi and climb-out. After the *Taxi* flare, Pete counted off the seconds aloud, released the brakes, advanced the throttles, and pulled out into our slot in the *parade's* growing line of slowly rolling aircraft. We were off to the races.

Heinkel HE-162 Volksjaeger (People's Fighter)

A *schwarm* (or flight) usually four, BF-109G's, Gustav's

5
The Mission

This mission began our part of Operation *Big Week,* which was a six-day bombing campaign to attack and destroy all Axis aviation-related industries in Germany and Occupied Europe. The overall aim of the mission was to reduce or severely limit aircraft and engine production to destroy or limit the enemy's air defense capability. In that aim, despite the destruction we caused, and the large aircraft losses we sustained, we fell far short. German aircraft production, was now widely dispersed in many very small facilities and factories and was continued even after the destruction of machine tools in the original factories by using hand labor by captured foreign workers and concentration camp inmates. The number of planes actually being built increased until the war's end.

An unexpected and unanticipated result of this onslaught, however, was the huge decimation of the German Air Force's day-fighter force by the massed fire power of the bomber fleets, not by fighter *kills.* So many of the German long-experienced and excellent fighter pilots were killed or incapacitated, that the Luftwaffe began a downward spiral from which it never really recovered, even though, despite the critical shortage of aircraft fuel, periodically German interceptors devastated large American air formations due to the lack of close air support by our fighter aircraft.

During the period in which I flew from the Fall of 1943 through the early months of 1944, our P-47 Thunderbolt fighters, even with their cardboard drop tanks, still did not carry enough fuel to get them past the German border and back safely to England. Yes, they really did use paper stiffened by resins to carry their extra fuel. The late-arriving P-51 "Mustang" aircraft were in too short supply to be of much help. Since then I've read of so-called escort missions during this period, but the real truth is written in the records of the high bomber losses during this same period.

Our 8th Air Force losses were commensurate with our opponents' losses, but with the vast aircraft production capability and the very large flying training programs in the US we could more quickly recover and promptly replace our losses in aircraft and crews. These six days of February 1944 were the 8th Air Force's greatest wartime losses in aircraft and personnel casualties, such as killed, captured and wounded in this momentous battle. The losses were more than in any other similar period during the air war. Historians have since called this the greatest air battle in history, primarily because of the very large number of aircraft engaged. However, for our 445th Bomber Group, these Operation *Big Week* losses ranked second only to the Kassel mission in September 1944 when the group was almost wiped out in a single day. After what appears to have been an ill-advised decision to leave the main bomber stream, the group was attacked and destroyed by a large group of German fighter aircraft, possibly as high as 150. Their fighter groups, or *Geschwaders,* normally of 100 planes, were much larger than ours - which I believe had 72 aircraft.

On our mission and in the proper order we taxied out of the hard stand and turned left, carefully following the bomber's running lights in front of us. We moved slowly and cautiously ahead like elephants at a circus parade, nose to tail, down and around the taxiway to the south end of the runway. A collision and the resulting fire with all the fuel and bombs aboard our

A rare photo of Luftwaffe ground crew loading 21cm Rocket onto an BF-110 Hornet. Normally four would be carried.

BF-110 Hornet

planes would have devastated our airfield and aircrews. We were taking off toward the north wind in rain and light fog. It was still as black as pitch. We could see only the runway lights and the plane ahead. We used only red lighting in the planes, since that interfered much less with our night vision.

There was no clearance by the control tower for each aircraft. Instead after the take-off flare was fired by the tower the first plane began its take-off roll. Radio calls would also have alerted those at German listening posts across the Channel and North Sea. The following planes counted off the separation in seconds and then began to roll. When the plane ahead of us began his take-off roll at the head of the runway, we turned right to line up with the runway centerline while counting off the seconds on the stopwatch that gave us separation later on the ground and in the air. The co-pilot reached over and set the runway heading into the gyro-compass for the instrument take-off. We rolled forward, gradually gained flying speed, broke ground, then climbed out flying the take-off heading for the briefed amount of time, exactly measured by stopwatch, then the pilot started a *standard rate* right turn to the *buncher radio beacon*, centered the needle, and homed in on the radio signal at the proper rate of climb and airspeed. As we passed over the beacon and the radio compass needle swung around we made another right turn to line up with the *splasher beacon* south of our runway. The co-pilot cranked in the new radio compass frequency. After we passed over that beacon we began a final right turn back to the runway heading - again flying a timed leg.

This triangular pattern allowed us to climb *on instruments* to the top of the clouds and hopefully avoid the other unseen planes from our own bomb group, who had taken off before us and those who followed us. Also, all around our airfield were many more nearby airfields launching their bomber fleets along their assigned routes. Add to this the relative inexperience of the very young pilots flying on instruments only and in very bad weather, and you had to wonder how it could have been done at all, especially on such a vast scale from the small island which makes up Great Britain.

The other four young officers, who had originally shared the Nissen hut with the four officers of our original crew, Lt Metcalf's crew, took off one morning in similar bad weather, but never joined the formation above the clouds as expected. They were never seen or heard from again. Since their wrecked plane was not found in England, we believed that the pilots had lost control while flying on instruments, *spun-in*, crashed in the nearby North Sea, and sank from sight. I'm certain that this was not unique. It probably happened many more times than we realized and added to our losses.

The aircraft instruments, radio compass, and the navigator's stopwatch were our only guides. Radar was not available back then even for lead crews, and even when it was installed later it was only available in the *lead* planes. Even when it was later installed, and after the operators were trained in its use, the *B-scope* radar of that day was a far-cry from anything recognizable as radar today. German radar, though far more advanced than that used by Britain, was used primarily for ground duty such as coastal defense and for aiming flak artillery. Their aircraft radar use was even more limited than ours. The night bombing over England by use of radio beams, was almost as imprecise as that of the Royal Air Force over the continent, because unlike laser beams, the radio beams broaden with distance and distort both the course to the target and the dropping signal.

As we climbed, the clouds around us over time gradually became lighter and then even visible, until we finally, no doubt with a great sigh of relief, broke out on top into the brilliant sunlight of dawn. Blinding. The vast number of B-24 Liberator bombers in sight all around us from our Second Air Division was both awesome and inspiring, but we were only a small part of the mighty air armada that would hit many German aviation-related targets that day. It's something we'll never see again in our lifetime, or even again in history, now that strategic bombing has changed so drastically in equipment capability, in weapons available, and in the tactics used. Enemy air defenses have so improved that no such air fleets could survive now. The Second Air Division was composed of 14 bomber groups, or 56 squadrons, arrayed around the city of Norwich in Norfolk County. It's in East Anglia, about eighty miles NNE of London. The 445th Bomber Group, to which our 703rd Squadron belonged, was based on an RAF station a few miles from Tibenham, 19 miles south of Norwich and near the LMS railway system which connected us to London and southern England.

Ahead of us we saw the our Libs with a large black *F* in a white circle on wings and twin vertical stabilizers. They were forming in a slow circle to the left. We turned just inside of the circle and slid across toward our position in the formation. It was very well choreographed. What had originally looked like a Chinese fire drill now became a well-ordered and massive formation. Even to this day, after all the years that have passed, and many more combat missions in later wars, I still marvel at the sights I saw in those days. These sights will always be with me. I wish that others could see them also and be as impressed as I was then

In front of our formation, with their *C* marked planes, was the 389th Bomb Group, veterans of many combat missions, from nearby Hethel. Their take-offs had begun ahead of ours and their group was already well formed. They were scheduled to lead us in the bomber stream to the target. We eased into line behind them and closed up to become part of the huge *wheel*. Next the great stream began a series of wide turns so that the leader's final right turn would pass over the coastal radio beacon at Cromer on the North Sea coast at the planned departure time.

Looking down I couldn't see the English coastline we were leaving due to the solid undercast covering it, but as we crossed the North Sea, to what we then called Holland, large breaks in the clouds below began to appear. I saw parts of the cold gray North Sea with its white wind-streaks streaming from the wave tops and then began to see the continental coastline. On the interphone the pilot told the gunners to *clear their guns* now that we were out over water. The gunners had to fire a short burst from each .50 caliber machine gun and check turret operation to make sure that all of our defensive firepower was in working order. Any equipment failure was an automatic *abort* and return alone to England. Our guns and their massed formation firepower was our only battle protection, even if, sometimes, a slim one.

Back then we had no actual fighter escort, regardless of what you've heard, only what was called "area cover". This information had been given to us as part of the overall pre-flight briefing. It told us what enemy airfields were being attacked along our route prior to our reaching the German border and it spelled out the probable enemy response from fighters and flak along our path and in the target areas.

Fighter aircraft, with some Spitfires and Hurricanes, but mostly P-47 Thunderbolts, would precede us. They would make strafing attacks against planes on the ground and try to catch those already in the air at German occupied airfields in Holland, Belgium and northern France, then would move on to those nearer to the German border. None of our fighters then had the flying range to accompany us into Germany itself. Early P-51 Mustangs joined in with them, doing the same types of attacks, but did not actually provide close escort to the bombers.

If the enemy fighters could get into the air soon enough, avoid the ground attacks, and climb to our altitude in time, they would make running attacks against the bombers formations and follow them across the border into Germany. There, the undamaged ones that were left would land at German airfields, refuel, re-arm, re-group, and try to resume their attacks again when we made our homeward flight. Sort of a *hail and farewell.*

Similar large fleets of B-17 bombers from the 1st and 3rd Air Division were also assembling in their formation areas. The 1st Air Division was as big or bigger than our 2nd Air Division, and was also composed of 14 bomber groups. They had arrived first in England and were already fully engaged in air combat while many of our bomber groups were still training. The 3rd Air Division with both B-17 and B-24 bomber groups was still somewhat smaller back then. Their bomber groups were still coming on line and would be doing so until almost the end of the European conflict.

The 1st, 2nd, and 3rd Air Divisions each had their own multiple target missions in other areas of Germany as we did that day. Targets were being attacked all over Europe at the same time by the bomber groups of all three air divisions of the 8th Air Force, and by bomber groups of the 15th Air Force's largely B-24 bomber groups from Italy. You can also see how Germany's air defense was so widely stretched across so large an area, and how it's response was limited by its constant fuel shortages The continuing air raids against the oil refineries at Ploesti in Romania continued to destroy fuel that Germany badly needed. Synthetic fuel production in Germany from lignite coal couldn't do much to meet the fuel needs either. On a previous mission the crew with which I flew had helped destroy one of these plants at Ludwigshafen. Bomber groups had all B-24's or B-17's - not mixed.

The Zuider Zee, the large shallow body of water in central Holland, was hazily visible. Fighters were called in from time to time, but neither our group nor the 389th ahead of us lost any planes before reaching the turning point at Dummer Lake just across the border in northwest Germany. In major flak areas such as this the German fighters stayed away from us and let the many Luftwaffe 88mm flak batteries pound away at us. With their well-trained and experienced gun crews, radar-directed accuracy, and massed batteries, they could and did cause great damage to many Allied bombers fleets.

This large lake was easily recognized and was an excellent place to get an accurate visual navigation *fix* in clear weather, but the 8th Air Force used it far too frequently as a turning point. It was pronounced *Doomer*, but the aircrews called it *Dumber*. It was easy to see that the people planning these missions weren't flying them.

The Germans had lined up a very large number of heavy flak guns. They had both the well-known, quick-firing, very accurate, 88- millimeter anti-aircraft cannons and the even far larger, though far fewer, and less well-known, 128-millimeter cannons. They could hurl their deadly projectiles far higher than our aircraft could fly. Here, in the close vicinity of Dummer Lake the large number of cannons quietly lay in wait to ambush our passage. Most were mounted on special railroad flat-cars for rapid movement and high mobility.

If the lake turning point was hidden by clouds beneath us, we jokingly said that the German gunners gave us a hand in our navigation by firing a large barrage of flak shells with heavy black smoke from their bursts above the clouds to help us identify the turning point.

As a matter of interest, the word *flak* is an acronym made up from three German words - *fl - a - k*, or *Flieger -Abwehr-Kanone*, which in English means *flyer defense cannon*. Many other words that we adopted from the Germans had similar parentage.

We continued with a turn to the southeast and a series of other turning points before reaching the Initial Point from which we began our bombing run. Fighters, both BF-109's and FW-190's, arrived in ever larger numbers and began head-on attacks with machine gun and 20mm cannon-fire. Our target was the Gothaer Waggonfabrik AG at Gotha, just west of Erfurt. The aircraft plant, 15 miles east of Gotha, was the primary source for the manufacture and assembly of twin-engine Messerschmitt aircraft that was so vital to Germany's war effort. It produced the BF-110, BF-210, and BF-410 planes in large numbers prior to our strike, but none afterwards since it was too badly damaged to even consider re-building it.

About midway to the target while on the bomb-run, the 389th Bomb Group, we followed in train, began a very slow turn to the left, away from the target. The sky was a beautiful blue, without a cloud in sight. The target's factory buildings and adjoining airfield were in open farm country and were unmistakable even covered in the beautiful, unmarked white snow. You couldn't mistake it. Our 445th group leader had to turn our 25-ship formation to the right to avoid the larger, straggling 389th formation, then turn left again to bomb the target.

This rather wild, last-minute, double turn, scattered our bomb group, but the lead bombardiers in each section were able to line up on their aiming points surprisingly well. Pete Abell, the pilot, opened the bomb bay's thin corrugated doors, which were built like a roll-top desk, while both he and I stood-by on the *Salvo*. His salvo was by switch, electrically, while mine in the nose was manually operated. My release was like an old Ford brake with a button and *Off* at the top, *Select* further down, and *Salvo* at the bottom position.

When our flight leader's smoke-cannister marked bombs fell clear of his bomb bay, we both hit our Salvos. It was a sight to see. First came the large white smoke markers attached to the lead aircrafts' bombs arcing forward and down and then our string of bombs chasing after them in a ballistic arc. As each of the four hundred, or so, 500-pound bombs from our Group exploded like punctuation marks on the white snow-covered area, with a bright red flash in its turn, the immediate area quickly disappeared under huge, black plumes of dirt, smoke, and debris. Strings of bomb impacts ripped through all the main buildings and then the massive fires erupted quickly with a large, rapidly climbing, pall of very black smoke. The main factory buildings and flight line hangars disappeared as they were demolished by bomb impacts and the following shock waves, even though it was a fairly large complex.

As soon as we were clear of the bomb run we began our turn to the right to the re-assembly of the defensive formation at the group rally point. I had an excellent view from the right side window that looked down in the turn. It's a scene I'll always see as clearly as a movie. Even though I've made many more such bomb runs later, both day and night, over enemy territory in later wars, none has been more spectacular.

On attacks such as this there were multiple aiming points, with so many planes assigned to each. The planes assigned to each aiming point were in close formation, but they were somewhat separated from the planes grouped on another aiming point. This caused some small diversions from the basic defensive flight formation, or *combat box*, but the scattering effect of the 389th's erratic left turn, our having to dodge their trailing planes, and the normal separation of aiming points, combined to cause us to be dangerously scattered.

This was not lost on the German fighter leader. Seeing our disarray he sent in wave after wave of single engine fighters. In most cases they passed by us at high speed above on the right, then when well ahead, began a diving turn to fire at us from head-on. This began to be a preferred German tactic to keep their interceptor aircraft in our field of fire as short a time as possible. Tail-end attacks, due to their slower rate of closure, leave the fighter pilots at the mercy of our gunfire much longer, but were necessary for the larger and slower rocket-carrying BF-110's and the smaller rockets of the FW-190's. With head-on attacks there was only time for one firing burst from the fighter and the bombers. After firing the fighters immediately rolled upside-down to put the armor plate under their cockpit floor toward the bomber's guns, while the enemy pilots dove under us and screamed past. As they passed beneath our formation, they rode their rudder pedals from side to side to throw off our aim. It always reminded me of looking down at fish in a stream. In actual fact, though, the under-seat armor of the German fighter aircraft was no match for the penetration and volume of fire from our .50 caliber machine guns.

Many of the attacking fighters were hit by the massed gunfire from our top and nose turrets, especially when we were in tight formation. We were able to put out a thick screen of heavy caliber bullets which would destroy engines and go all the way through the enemy fighters regardless of their thin armor. We saw many put out trails of smoke and flames, and some even exploded into fireballs which no one would have survived. For the most part, the fighters had to attack singly against each section of our formations due to their extremely high rate of closure to prevent colliding with each other. The combined rate of closure was almost 600 plus miles per hour. This single plane attack exposed each of them to many more guns. An expression that popped into my mind back then was, "like trying to run naked through a firing squad".

Some enemy aircraft when hit collided with bombers when the pilots were killed or seriously wounded. Parts of enemy planes also hit our aircraft when fighter planes exploded or were badly damaged. In this battle, part of a damaged fighter flying past tore into our left wing and created visible damage, although we didn't seem to be adversely affected.

The German fighters, depending on type, had either heavy or light machine guns, but each also had at least two 20mm cannons. The cannons could hit us at a greater distance than our .50 caliber gun's range could reach them, but we had a far greater volume of fire and better chance of hitting a target. Attacks took only tiny instants of time, but repeatedly they created a hectic scene with the gunners screaming about battle damage, fighters coming in, going down, or exploding in red flashes of fire and smoke. Five minutes seem like a lifetime when waiting, but here you weren't even aware of time. I tried to record information about our group's planes getting hit, burning, or going down. I tried, too, to keep score of enemy planes hit or destroyed by scribbling notes and times on my map. I guess it was like trying to describe an auto accident while it was happening. Voice recorders, though not invented then, would have been very useful.

Having later seen many film clips photographed by German fighter cameras during attacks against our planes, I have seen only those that were made by rear attacks, but none that were made during head-on attacks. If there are any such films, they would show the difference in the time span or slow rate of closure of tail attacks versus the vastly faster closure of the nose attacks.

When possible we also tried to use our small, hand-held, 5" x 5" bomb cameras to record what happened with our damaged planes, to see who got out as the aircraft fell, in the hope that the film when developed would show crew members jumping safely from spinning or burning bombers.

I learned much more later from British historian, Charles Whitings's book about this battle, that German Junkers JU-88 *Schnellbombers* (fast bombers) had dropped small parachute fragmentation bombs and dragged long steel cables to entangle our propellers to break up our formation, though I didn't see the cables. During the battle we did see the small parachutes mines that were called out.

We'd gotten back into formation with several other planes from our bomb group and were heading to the southwest to join the bomber stream's homeward route, when more and more fighters began attacking us. We had many bullet holes and other damage in the nose area through which I could see bright sunlight, but even with many bullet holes, apparently there was no major damage. Most I hadn't even been aware of when they happened. The nose turret fired off and on, shaking the whole nose. Someone called, "BF-109, one o'clock high." I looked up and forward through the big glass astrodome and saw the plane growing bigger, then his gun flashes along the leading edge of the wing were clearly visible. The glass disappeared into brilliant dust as the bullets sprayed above me and then blew out the pilot's left windshield.

Later I found out that the bullet stream which barely missed me had hit the pilot, Pete Abell's helmet. The first few rounds knocked off the small leather-covered Wilkinson steel helmet and then blew out the right radio headset from his leather flying helmet underneath. As his head involuntarily spun to the right, another bullet took off his right eyebrow. It didn't damage his skull, but left a bloody furrow where the right eyebrow had been.

This was earlier in the war, before the GI steel helmet was modified with steel ear flaps for aircrew flak use. We had been issued the RAF's Wilkinson helmet with leather covered thin steel plates. It looked more like a football helmet. It fitted over our leather flying helmet and was useful against small flak pieces, but not bullets like the German 7.92mm light aircraft machine gun or anything heavier.

Neither helmet, the British or American improvised one, was very effective, but we needed something light enough to use that would allow us to do our jobs and give us some small amount of flak protection, especially against the 20mm cannon shells that exploded when passing through the aircraft's skin and sprayed metal shards around us.

Since by then we were already getting mostly head-on attacks by the enemy fighters, we needed the best and most experienced turret gunner in front. This is why the normal tail turret gunner, Merriam, was placed in the nose. Reeves, normally the right waist gunner, took his place in the tail turret. Also because we were not leading a section, or even deputy lead, we didn't need a bombardier in the nose. As navigator, I could toggle out the bombs or salvo them as required. That's why Lt Chuck Stealey, the assigned bombardier, was being used on this trip as the right waist gunner.

While this may seem unusual now, it wasn't then. In command aircraft, also, the co-pilot, who was checked-out on turret gunnery, often manned the tail turret so he could pass formation information to the air commander who rode in the co-pilot's normal seat. 8th Air Force orders were that pilots, or crew commanders, were to fly the left pilot's seat at all times, not the combat leader.

Somewhere in this time frame, Staff Sergeant Lonnie Reaves in the tail turret had both legs smashed by gunfire from a rear attack, and in all probability had quickly died from blood loss or shock without ever calling out to the pilot or other crew members. In the heat of battle, his voice was not missed soon enough. Lt Chuck Stealey, at the right waist gun position, had a pair of 20mm shells explode around him after passing through the aircraft skin. The one that detonated in front of his face blew off his aircraft goggles and left unburned black gun powder particles embedded in the skin under his eyes. The second one completely shredded the seat of the G I olive drab wool pants he wore over his blue bunny heated flying suit. In both cases he was not hit by the shell fragments, or shrapnel, that flew past him. He was probably very startled but not hurt.

Soon after this a huge explosion took place. Everything shook violently as if the plane had exploded. I had no idea what had happened. Probably no one else knew either. Beneath me I could now see snow and a line of trees far below where I'd formerly seen only the floor of the front compartment. I now saw a large hole. With hands holding on to either side of the fold-down navigation table and with my feet braced wide apart at the bases of the large turret ammunition boxes on either side, I kept my balance during the turmoil. The aluminum skin of the plane was gone under me. Only some of the ribs and bracing were left.

There was an immense roaring of wind racing by the plane past the missing floor, around a side window with the glass missing, and through the blown out astrodome. Pete told me later that we'd been hit by an aerial mine dropped by parachute from twin-engine fighters above us, because we could see small parachutes falling around us, but I didn't learn the actual true story for many years afterwards.

I did not find what had occurred until a German historian friend, Horst Jeckel, dug up the name of the German BF-110 pilot who'd been credited with shooting us down. The weapon he used with a timed-fuse was a large ballistic, but not guided, rocket that he had fired at our formation. Fortunately it had exploded some distance under us and not on impact. This was the source of the very loud explosion we'd heard, and it was the weapon that had blown much of the lower metal skin off the plane. It also destroyed many of our aircraft systems and had set the aircraft on fire, probably by rupturing wing tanks.

Much more recently, I learned from a nearby friend, Sam Mastrogiacomo, then a staff sergeant tail gunner, now a retired master sergeant, more of what actually happened. His crew, from the 702nd squadron in our 445th Bomb Group, were in a nearby B-24 behind us during the battle. His plane was approached from the rear by a Messerschmitt twin-engine BF-110. He fired his tail turret guns to hit or drive off the enemy. The fighter slowed, slid back out of Sam's firing range, and then launched two large rockets, one from under either wing. He said he saw them coming toward his plane and thought it was the end for him and for all of his crew. But his luck held, the rockets arced under his plane, and passed safely underneath. His plane, though damaged by gunfire, made it safely back to our base in England. In all likelihood, one of the two rockets that passed under his turret was the one that exploded under our plane. We weren't hit by one of the parachute mines Pete Abell was dodging.

Everybody in the back of the plane had jumped out when the fire roared back to them. They didn't wait for an order, fortunately, because none was possible. There was no interphone or alarm bell bail-out warning left. I looked up toward the cockpit and saw Pete waving to come out of the nose. I moved forward carefully and pounded on the nose turret door and waved the gunner back towards me. I pointed out the missing floor, because the noise level from the roaring air was too high for talk. We both crawled back over the debris, the open hole, and through the tunnel passage under the main deck floor. The gunner went first. Staff Sergeant Bob Merriam, snapped on his chest pack parachute and dove out the right side of the bomb-bay whose roof was a mass of flames being driven back through the plane by the wind from the open bomb-bay doors and the wind from the nose.

Pete Abell was on the right side of the catwalk fastening the leg straps of his seat-pack parachute before he tumbled out into space. Husky 2nd Lt Bernard Anderson, the co-pilot, was on the left side fastening his chute when I hastily snapped on my chest-pack chute and dove over him through the open bomb-bay door. I didn't hit him or the bomb-bay wall as I tumbled out into the slipstream on the left side. Our burning plane, engulfed in smoke and flames went only five miles or more after we got out before it exploded into a fireball and spread debris all across a snow-covered farm field far below. The body of tail-gunner, Lonnie Reeves, was later found in a pile of aircraft debris on the ground. The crash wasn't far from where I landed, but I don't remember seeing or hearing it. The plane exploded over the village of Herbstein.

We'd been instructed to hold off our parachute opening release until fairly close to the ground so we would not be spotted as soon by the enemy police and capture crews, but I'm sure I pulled the ripcord far too soon, perhaps at seventeen or eighteen thousand feet, almost as a gut reaction. I felt a great sense of relief and a huge peaceful silence as the roar of the bombers and fighters rapidly left the area. I don't even remember feeling cold, even though the temperature was far below zero. When I neared the ground I selected a forest clearing for a landing spot on the side of a hill not far from a red-roofed village. I tried to control my wind drift to steer to it by using the parachute shroud lines. When I neared the ground I suddenly noticed that I was drifting backwards rapidly. Then I slammed into the deep snow and a hit a very hard surface which I later thought was a low wall. It was probably a fallen tree. While sitting I unbuckled the parachute harness and pushed it off my shoulders. I had great pain in both legs. Seeing a group of men lead by a man in a green uniform climbing up the hill through the snow towards me, I tried to get up and move, but after several very painful, lurching steps I again collapsed into the deep snow and waited for the group approaching.

As a follow-up I learned long afterwards that the 389th Bomb Group's lead crew had suffered an oxygen system failure at the last moment on the bomb run. The pilots had slumped over and caused the slow left turn on the bomb run which caused so much turmoil with aircraft scattering and havoc during our bomb run. The pilots and bombardier did revive in time to lead the 389th Bomb Group across the target. Their groups's bombs added to the total destruction caused.

For this combat mission's complete success, its great value toward the victory against Germany, and for the very great penalty we'd paid in the number of airmen killed, wounded, and taken prisoner, the President awarded the 445th Bomber Group with the Presidential Unit Citation signed by General George C Marshall.

Many years later, I learned, through the research of German historian, Horst Jeckel, that we were shot down by pilot and Feldwebel Joseph Holtzmann when he fired a 21 cm aircraft rocket at our formation. The rocket missed the aircraft of a friend from the 702nd Squadron of our Group, now retired Master Sergeant Sam Mastrogiacomo, when it flew under his aircraft safely and exploded under our plane. The time was 1340 hours on the 24th of February 1944; the location was southwest of Fulda. Feldwebel Holtzmann was with the 1st Squadron of Zerstorergeschwader 26 (Destroyer Wing 26), flying the twin engine BF-110 Zerstorer.

Another note: My youngest brother, Jim Reus, a navy veteran of WWII, later married a female medical doctor who served with the Veteran's Administration at a hospital in Maryland. This woman had been a high school senior attending class fifteen miles away in the town of Gotha at the time of this air raid. She vividly remembered what had occurred that afternoon having heard the large explosions and seen the enormous smoke pall. There was no damage in the town of Gotha, because of the distance to the factory from town.

She was from a *Volksdeutsche* (or German ethnic group, such as that in the Sudentenland German group in northern Czechoslovakia, or the Russian group called the Don River Germans), family in Estonia. They had been evacuated to Germany to protect them from the Russian occupation that came about with the Russo-German peace pact and the partition of Poland that took place just before WWII began. She told us that she had seen the destruction at the factory we'd bombed and that the entire complex was so heavily damaged that it was never rebuilt.

We also have many *Volksdeutsche* groups in the United States. One such is New Braunfelds (or New Brownfields) located between Austin and San Antonio, Texas. After the Civil War, a German nobleman immigrated to the United States with a sizable number of his people to central Texas to colonize a new land. Here, and in other such areas, they moved in a group, unlike many of our ancestors who came over as individuals or as family groups. There are many others, of course, but most are located in more northern and cooler areas of the US, such as Wisconsin, Minnesota, North Dakota, etc.

6
Distinguished Unit Badge

GENERAL ORDERS)
NO. 42)

WAR DEPARTMENT
WASHINGTON 25, D.C., 24 MAY 194

DISTINGUISHED UNIT BADGE

* * * * * *

BATTLE HONORS--Citation of units -- -- -- -- -- -- -- -- -- -- -- -- -- -- -- -- -- -- --Section IX

6. As authorized by Executive Order 9396 (sec. 1, WD Bul. 22, 1943). superseding Executive Orde 9075 (sec. III, WD Bul. 11, 1942), citation of the following unit in General Orders 129, Headquarters 2n Air Division, 11 April 1945, as approved by the Commanding General, European Theatre of Operations, i confirmed under the provisions of section IV, WD Circular 333, 1943, in the name of the President of th United States as public evidence of deserved honor and distinction. The citation reads as follows:

The 445th Bombardment Group (H), 2nd Air Division, is cited for outstanding performance o duty in action against the enemy. On 24 February 1944 this group participated with other heav bombardment groups of the 2nd Bombardment Division, in an attack on the Gothaer Waggonfabrik, A.G located at Gotha, Germany. On this occasion the attacking bombers met and overcame the fiercest an most determined resistance the enemy was able to muster in defense of this target, so vital to his ability t wage war. Unprotected by friendly fighter cover the 445th Bombardment Group (H) was under almos continuous attack from enemy aircraft for a period of 2 hours and 20 minutes. Although antiaircraft fir was hurled at the formation along the route to and from the target as well as at the target itself, the mos deadly opposition was given by enemy aircraft. For 1 hour and 20 minutes before "bombs away" savag attacks were made by single and twin-engined enemy fighters in a vain attempt to keep the bombers fron reaching their target. On the actual bombing run, the critical period of each bombardment mission, fierc and relentless attacks were unable to keep the bombers from accomplishing their task. For another hou after bombing the group continued to be the object of ferocious fighter attacks. Of this group's 25 aircraf which penetrated enemy territory, 13 were lost to these fierce fighter attacks, which number i approximately twice the loss suffered by any of the other groups participating in this mission. In addition, 9 of the surviving 12 aircraft returned from the mission with battle damage. With heroic determination th 445th Bombardment Group (H) flew its assigned course, destroying 21 enemy attackers, probabl destroying 2 more, and damaging 7 during the long running battle. The target was located and bombe with extreme accuracy and devasting results. This target, the most important source of ME-110's, was so well hit that the enemy air force suffered a most telling blow. The courage, zeal, and perseverance show by the crew members of the 445th Bombardment Group (H), 2nd Air Division, on this occasion were i accordance with the highest traditions of the military service of the United States and reflect great credit o themselves and the group, and the Army Air Forces.

BY ORDER OF THE SECRETARY OF WAR;

G. C. MARSHALL
Chief Of Staff

OFFICIAL:
J. A. ULIO
MAJOR GENERAL
THE ADJUTANT GENERAL

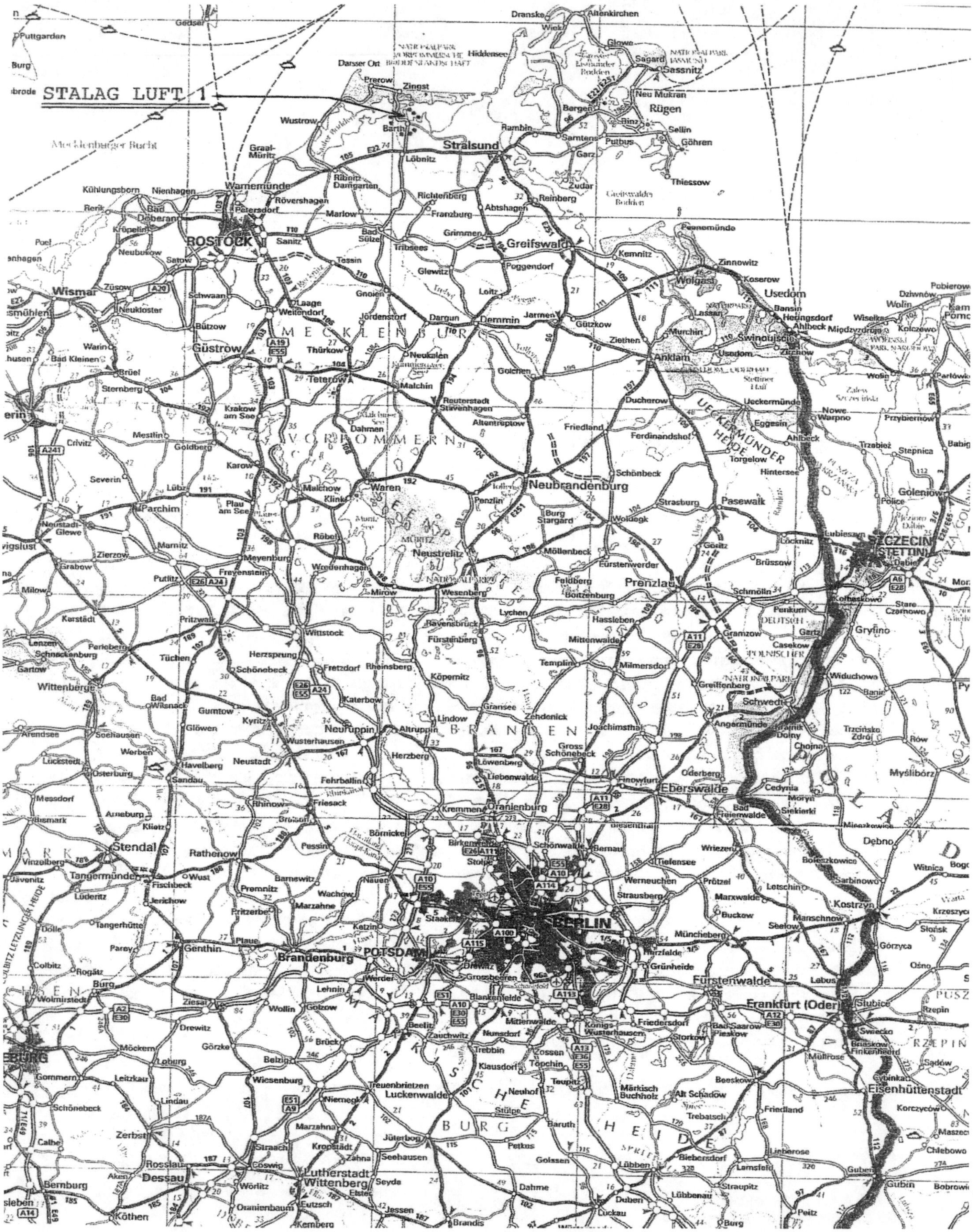

Map- Stalag Luft One on the Baltic Sea

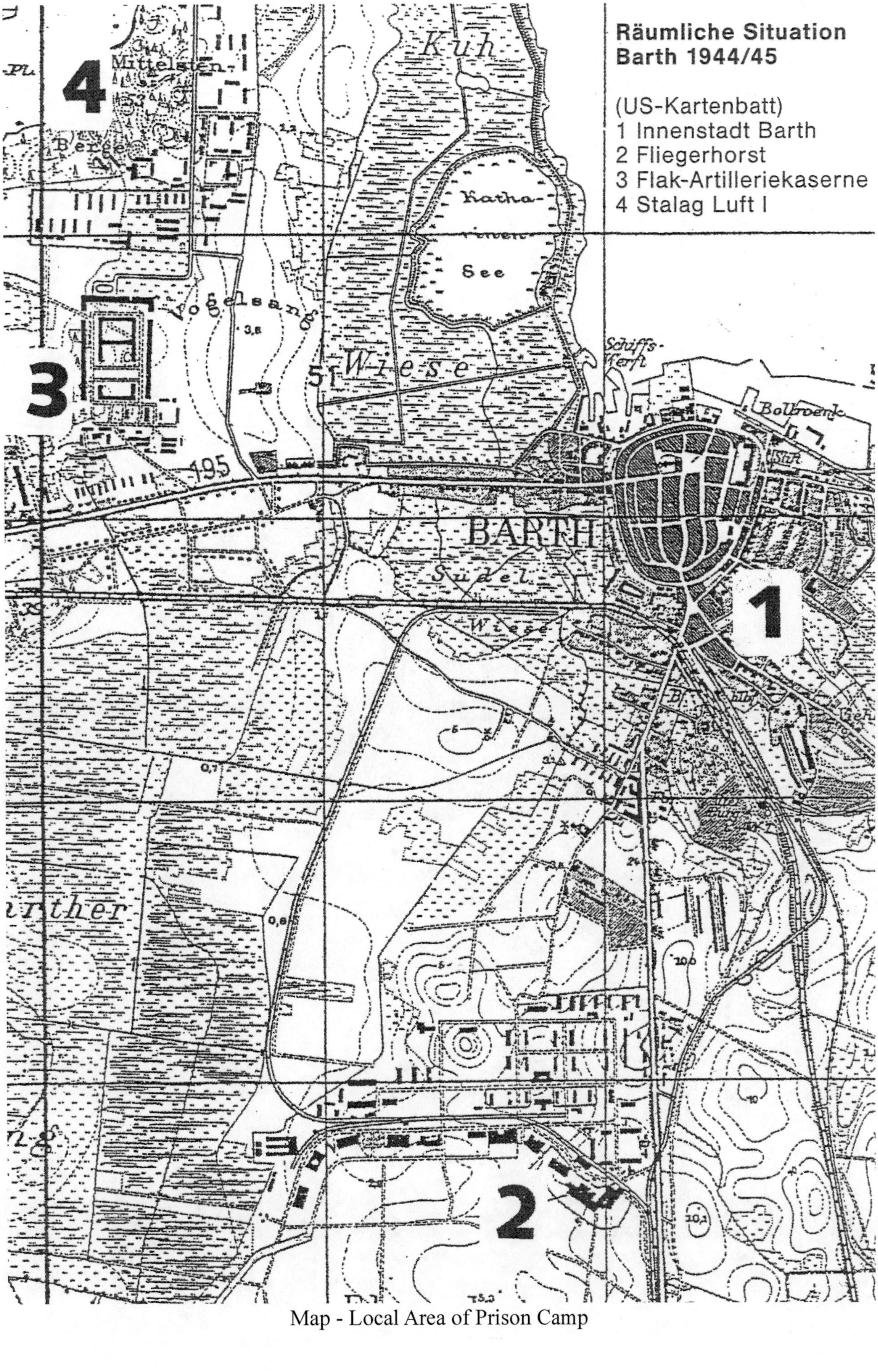

Map - Local Area of Prison Camp

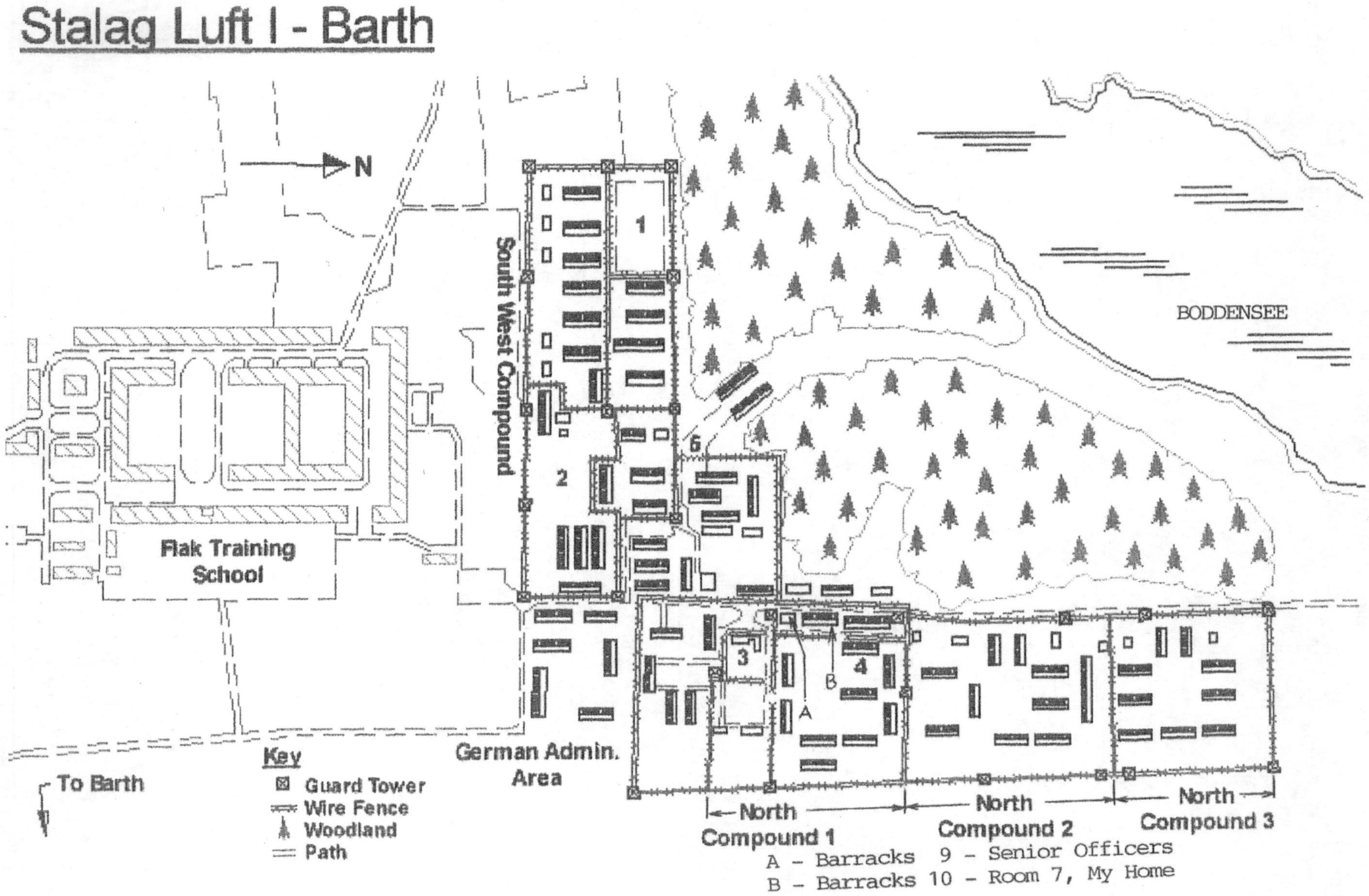

Map- Stalag Luft One Camp Layout

Crew Members On Gotha Mission
Operation *BIG WEEK*, 24 February 1944

Pilot	1st Lt Peter T Abell	POW	Engineer / Top Gunner	M/Sgt Donald T Caton	POW
Co-Pilot	2ndLt Bernard I Anderson	POW	Radio Operator	T/Sgt Donald E Watts	POW
Navigator	2nd Lt Joseph H Reus	POW	Nose Gunner	S/Sgt Robert W Merriam	POW
Bombardier	2nd Lt Charles T Stealey	POW	Waist Gunner	S/Sgt Allen S Alford	POW
Tail Gunner	S/Sgt Lonnie H Reeves	KIA			

*NOTE: The four officers were sent to Stalag Luft 1, an officer's camp.
The NCO's were sent to a non- commisioned officers camp.

Stalag Roommates
Room #7, Barracks 10, North One Compound, Stalag Luft 1
Normally 12 men in double bunks in 12 x 15 foot room

1. 2nd Lt Kenneth Covey	7. 2nd Lt Joseph H Reus
2. 2nd Lt James Ellis	8. 1st Lt Michael O'Shea
3. 2nd Lt George Kilduff	9. 2nd Lt Arthur W Starratt
4. 2nd Lt Leonard R Lovelace	10. 2nd Lt Charles T Stealey
5. 2nd Lt "Mack" McConnell	11. 2nd Lt Steiner
6. 2nd Lt Earl Rathke	12. No Record

Roomates

Prisoner of War
Certificate of Captivity

JOSEPH H. REUS
Serial Number A0-687962
2nd Lt, US Army Air Corps
8 AF 703 BS 445 BG
2 Combat Wing 2 Air Division

POW # 2993
Stalag Luft Barth
Barracks 10, Room 7
North One Compound

American Ex-Prisoners of War
Arlington, Texas

POW Certificate of Capture

Kreisleiter Zürtz of Lauterbach with Gestapo Untersturmführer (2nd Lt)
(County Chairmam under NSDAP Nazi Government)
Photo Credit: Fotoclub Lauterbach e.V. Negative No. 53/24

GERMAN AIRCRAFT AT GOTHA
8th Air Force Operation *Big Week*

Although the defenders did not arrive early enough to prevent the total destruction of the Gothaer Waggonfabrik A.G. aircraft factory at Gotha, Germany, they did destroy many of the American bombers as they turned homeward. Our 445th Bomber Group from Tibenham, England, lost thirteen of the twenty five B-24's that helped make the very successful attack. This huge aircraft plant was the largest and primary producer of twin engine *Zerstorer* (destroyer) aircraft such as the BF-110, 210, and 410's. We called them Messerschmitts, but in fact they were not *ME's*, but were *BF* for *Bayerisheflugzeugwerke*, (Bavarian Aircraft Works) followed by model number.

Defenders taking part were single-engine BF-109 and Focke-Wulf -190 fighters, BF-110's with and without the 21 centimeter aircraft rockets that brought us down, BF- 210's and lastly BF- 410's with multiple 20 millimeter rapid-fire cannons. Flying above us were Junkers JU-88 light bombers dropping parachute mines that exploded on contact. I've heard rumors of other participating defenders, but I have no direct knowledge of them. It was a very exciting and noisy time while it lasted. After I bailed-out, German and American planes raced away while I hung from my descending parachute in a sudden and vast silence.

Gruß aus Angersbach or Greetings from Angersbach
Views of the village.

Common railcar, like the English *goods wagon*, in which we spent four days traveling from Oberusal to Barth and Stalag Luft 1. Twelve prisoners and four guards were in each car.

7
Capture

As I sat in the hollow in the deep snow circle around me that I had just made while attempting to get onto my feet, I watched a party of six or eight Germans climbing up the steep slope through the snow drifts from the small road or footpath below. I had no thought of trying to escape, because my earlier attempt, after a few lurching steps, had convinced me that the pain in both ankles from the hard landing was too severe to allow walking, running, or even standing unassisted.

The leader, tall, slim, and wearing a long, green, military-type coat with a green kepi-style police cap, was overtaken and passed by two of the younger men moving up the hill more quickly, but as he approached my location the leader shouldered his way to the front. He had an MP-40 machine pistol slung over his shoulder from a strap. I didn't notice then, nor now remember, if any other members wore a uniform or were armed. My impression is that the group other than its leader were in heavy civilian winter clothes. My attention was centered on the leader, probably because of his impressive uniform and leadership bearing. Being young and totally inexperienced as we all were then, I thought of the approaching men as people coming to lend their help , and not as dangerous enemies.

When the leader approached he motioned with his hands to get up. Being still in a complete daze from my recent hair-raising adventures and the hard landing shock, I couldn't think of anything to say in English, much less use the few words of German that I still retained from my German grandparents. Instead I told him I was "Leutnant Joseph Reus," and pointed to my legs.

The uniformed leader, I later found out, was Polizei Meister (police chief) Schaller of Lauterbach, the nearby county seat. He reached out with both hands to lift me, then put his arm around me and supported me as we both slowly made our way down the fairly steep hill to his small vehicle below. I believe it was a *Kubelwagen*, or German Jeep. He helped me into the right seat, walked around to the driver's side, and said a few words to his men before driving forward.

One of the men had gathered up my parachute and its harness to bring back to town. Many years later when I visited the area someone told me that the chute had been cut-up to make a wedding dress for a wartime marriage. I hope it served the bride on such an important day as well as it did me and I wish the couple well, even though I was then somewhat the worse for wear.

The chief drove eastward, away from the direction to Lauterbach. Unaccompanied by the other men, we drove a short distance to a small, narrow wooden building just off the road on the right.

The chief helped me from the vehicle and inside, sat me on a long bench, then sat beside me. There were several people already there, men and women. It seemed to be some sort of farmer's meeting house. Time dragged by while no one said anything. They just stared, but apparently, out of curiosity, not anger.

Finally, after what seemed like half an hour, the door opened and a young man, probably a youth in his teens, came in accompanied by an older man. After he said hello to me and began speaking hesitantly in English, I began to get the idea that he was probably a language student, who came as an interpreter. The chief spoke to him in German, then the lad attempted to question me in English. I made a pretense of trying to help him by seeming to be gentle and helpful, but my real purpose was to defeat his attempts at communication by confusing him. It wasn't difficult. I gave him too many similar sounding words under the guise of apparently helping him. This went on for some time as the lad became more confused. Finally the chief could see that nothing was being accomplished. He waved off the lad and assisted me back to the Jeep for the short ride westward to Lauterbach.

We arrived at a large building with several wide stone steps up which I was partially carried. After we entered the building I was taken to a staircase on the left with white walls on either side that led downwards. At the bottom was a short corridor brightened by a window with fading sunlight. To the right were three cells with solid wooden doors with *Judas windows*, or covered peepholes, then an end wall. I was taken into the nearest, or first cell, and helped onto a bunk before the door was closed.

I laid back and closed my eyes, but after a few minutes the door re-opened and a man in a suit entered carrying a medical kit and said he was a doctor. He examined my ankles and legs, looked at the few small cuts, abrasions, and several minor flak hits that had little metal penetration. He rubbed each with cotton and alcohol, then told me in a few English words that I didn't require hospital care and my ankles would recover in a few days. Then he shook hands and left the cell.

I lay back and dozed until I was awakened by the opening of the cell door again. Another man, probably in his early forties, wearing a suit, came in, smiled, and sat on the foot of the bunk, since there was no chair. He mentioned a name which I didn't get, but for some reason I had the impression that he was the mayor or a city official. He chatted on for some time in fairly accurate English, but asked no questions. He said he'd grown up in America, on a farm in Iowa, and that he'd returned to Germany with his family years before. I listened, but didn't speak, only nodding now and then. Finally he shook hands and left.

Later, at twilight, a policeman in a dark uniform brought dinner. It was in a group of five nested, round, enameled metal dishes with a u-shaped metal carrying handle. He showed me how to separate the dishes. Then he really astounded me by handing me a knife, fork, and spoon wrapped in a plain white cloth napkin. I felt more like a guest than a prisoner. The five metal dishes held barley soup with carrots, boiled cabbage, two fried potato cakes, two small slices of buttered white bread, and apple sauce for dessert.

Even writing this, I still find it hard to believe the sequence of events that took place. I don't know what I would have expected, because I'd never thought ahead as far as being shot down again, much less of being a prisoner. I don't even remember ever speculating about such a situation, but if I had, I don't believe I'd have expected to be received this well. It was more like being an unexpected guest rather than a prisoner of war. My only interest afterwards was in dozing after a very long and exciting day that started the night before, not in wondering what the future would bring. I was awakened by the cell door being opened a third time.

My next visitors, in the early evening, were two Luftwaffe NCO's. Evidently they had been informed of my ankle injuries, because they jointly picked me up under each arm and carried me back up the stairs to the main floor and then into an office with a large desk. I was put into a chair in front of it. Across from me at the desk was the same man in a suit who had talked to me in the cell, Chief Schaller in police uniform, and then an older, white-haired man in a black uniform with silver insignia, including SS-runes. There were four square *pips* on his left collar tab. We hadn't been taught about enemy uniforms or ranks, but I later learned that he was a *Sturmbannführer*, or major, and on the staff of the *Gauleiter*, (district leader) as the Gestapo chief of the area.

The major was the only one who spoke to me. He asked me in accented English who I was. I answered, "2nd Lieutenant Joseph Henry Reus, A0-687962." He asked for my ID tags, and for the first time I realized I didn't have any. Somehow the tags had been mislaid or lost during my earlier hospital stays just prior to this mission. Chief Schaller slid a pad and pen across to me and made a writing motion. I wrote down the same information on the note. Amazingly enough this note is still located in the police files at Lauterbach, and is also shown, in this book, *Luftkrieg uber dem Vogelsberg*, (Air War Over The Volgelsberg), written by German historian friend Robert Keller.

Next the major handed me a Red Cross form printed in German and in English and told me to fill it out so my family could be notified where I was. When I finished writing and handed back the pen and completed form, the major asked, "What unit are you from?" I said, "I'm sorry, sir, that's military information." He looked up, pounded the desk, and answered, "Good German boy." Again I was very surprised, probably shocked, to put it mildly. No one had pressured me or insisted on answers. None of this was dramatic, if anything, it was almost sedate. The major after a few more words in German to the men at the table, waved the two NCO's forward from the wall where they had stood behind my chair. They lifted me under the arms and returned me downstairs to the cell and bunk.

Next morning I was awakened before daylight by the same policeman who'd brought dinner. He had brought two slices of buttered black bread and a big mug of what I later learned was *ersatz* coffee. It was strong and black, but sweetened. It wasn't really like coffee, but as I later learned at Stalag Luft I, it was made from acorns, but was better than nothing, even if black.

At dawn the same two NCO's returned. They helped me outside to their small light-gray Opel sedan, probably a Luftwaffe vehicle. The older and senior man spoke fairly decent English, but said only a few words as we drove several miles through the snow-covered countryside and then out into a large field literally covered in all directions by a great number of shattered aircraft parts. Most were fire-blackened and ripped into jagged shards. None were as big as a car. When the senior NCO asked me if this was my plane, I told him that I had no idea whose it was or even what it was. I had seen a piece of corrugated bomb-bay door, so I knew it was the wreckage of a B-24, probably mine, but I didn't want to help him in any way. The twisting and bending caused by a violent explosion was very evident when I saw .50 caliber bullets that were bent at right angles without having exploded. (It's the war.)

Many years later, on return to the same field with historians Horst Jeckel, Robert Keller, and several others, including a TV crew from Frankfurt, we met the field's friendly owner and his wife. As a very young man on his father's farm during WWII he had transported the four, heavy, Pratt & Whitney, 18 cylinder, model 1830 engines, using a very large farm wagon into the town of Lauterbach for the German scrap metal drive. I can't imagine how he lifted the heavy engines and other larger parts. On my return to that same field in 2006 we used an electronic metal detector supplied by Horst Jeckel, and even after sixty-two years of farming, we still found more small parts or scraps of metal beneath the dirt.

I was also told during the 2006 visit that I was fortunate that when I was originally captured, the actual mayor of Lauterbach then was away doing military service - and that as Kriesleiter (literally circle leader)or county chairman for the Nazi government, the NSDAP, I would not have been treated so well. He was an ardent adherent of Hitler. Also I learned that the local Gestapo leader on the county staff was an SS Untersturmführer (SS second lieutenant). He, too, was away on military training. The fellow who chatted in my cell was actually the deputy mayor who was not as political.

Back then, at the time of capture, after we drove from the wreck scene we traveled through fairly deep snow on small farm roads to larger roads that led to a major highway that carried us to Giessen. There we came to a series of large buildings enclosed by high, barbed-wire topped fences. I later learned that this complex was a Luftwaffe detention center, or prison, for "retraining" its own personnel.

I was signed in, then taken to a large holding cell with ten or more American airmen. I supposed it was yesterday's *catch-of-the-day*. I recognized Pete Abell, limped over and sat down next to him. He looked much better than when I'd last seen him. Then he'd had dried blood caked down the right side of his face from the bullet that had taken off his right eyebrow. I asked if all the men were Americans. He said, "I don't really know for sure." We talked quietly in generalities. Supper was a bowl of soup, a slice of black bread, and more ersatz coffee.

One of our guards, a young disciplinary prisoner himself, probably a *trustee*, spoke some English. He hung around the barred wall that separated our large cell from the corridor. He told us how, "Chermany vas going to vin da vor vit dair ooh-boats." He carried on with more enlightening "information" that raised our spirits quite a bit. It wasn't as good as a movie or much of an act, but was somewhat comical and helped use up the time until we fell asleep - even with all of the lights left on. None of the American airmen bothered to answer him. We guessed that he was trying to convince himself rather than us.

The next morning we were awakened early with *brot und ersatz*, then waited around until we were taken by truck to the railroad station at Giessen. On arrival we were loaded into a passenger railroad car for the trip down to Frankfurt. Wooden seats for the lower classes. There were about fifty Americans in the group, with a small group of Luftwaffe guards at either end. The civilian passengers were crowded forward to another rail car.

Our eyes were glued to the windows as we passed a heavily damaged airport, probably the one we now call Rhein-Main, or Frankfurt airport, as if we were merely on a tour.

When the train backed into the Hauptbahnhof, or main railroad station at Frankfurt, we saw the enormous damage the stately old train station had received from many bombing raids, both by the Americans and the British Royal Air Force. Only the eastern track, the one we were on, was in operation or usable. It was probably kept in a barely usable condition with salvaged material, and had to be repaired over and over, so that this part of Frankfurt wouldn't be cut-off from the rest of Germany. All of the eight or ten tracks on either side of the various concourses, or platforms, to our left, or westward seemed to be completely abandoned.

There were gray, or blackened, twisted and broken, metal beams of all sizes and shapes, enough glass fragments to give the impression of a crystalline ice-storm, and shattered concrete dust, blocks, and pieces or all sizes, shapes, and descriptions piled across the other tracks in some places higher than a house. If there were actual bomb holes they were hidden by the masses of fallen material. What had been the high, arched, glass roof supported by metal arches was now outlined only by its absence.

All of this was permeated by the heavy, burnt smell of a disastrous fire. It was probably a shock for each of us to see, for the first time, and to be actually in the midst of, what our bombing actually caused. Our high perspective and quick passage had not given us this sense of desolation. Probably one or more men in our Pow shipment had previously bombed Frankfurt - perhaps even this great railroad station. We were now able to experience these scenes of almost utter destruction many times as we later traveled across Germany. It's a lasting and devastating memory. Our group was quiet and somber as we realized what war had brought to the ordinary German people.

On a later visit to this same beautiful old station with a group of former POW's who came through this terminal during WWII we saw that it has been rebuilt to its former turn-of-the-century grandeur. Now everything is back in place and it doesn't look as if even a pane of glass had ever been broken. The German reconstructive methods, both here and at other places we've visited, such as Dresden, are both outstanding and amazing. Nothing looks new or replaced - it looks as if it has always been.

With two friends, including Pete Abell, who supported me on either side, we left the train for the platform, and then moved to the station's wide platform ticketing area . When we got there, four or five elderly ladies in gray and red caps and dresses with Red Cross armbands were ladling out macaroni soup from a large deep pot into bowls and handing it out to us with slices of black bread. Our guards were being fed too. We were very grateful. The macaroni was whole wheat and unbleached - unlike ours, it was a spotted gray, but it tasted to us better than caviar would to a gourmet.

While I was sitting on the floor eating, a young German army captain in infantry field gear and a heavy rucksack tapped me on the shoulder and asked me in good English who we were. I told him we were captured American airmen. He told me, and a whole group of quickly interested Pow's, that he was heading back to his unit on the East Front after a short home leave. Unfortunately, before we could say much more, he was ushered on his way by nearby police. I think we all wished him good luck and certainly wished that we could have talked to him much longer. I don't know what the rest of our group thought of this young man, but I was not alone in being more sympathetic to him than to our joint enemy, the Russians, who were mistakenly believed to be our Allies. He was heading to a far worse place than we were - and with a far, far lower chance of survival.

When we finished eating, we were again lined up and taken across the deep lobby and out the nearby large main door to the street. Traffic was stopped along the wide street and we were led across toward a long tandem streetcar or two-unit tram with wooden sides like old Baltimore streetcars.

As we crossed toward it, I spotted a newly-fallen, large, shiny piece of an 88mm flak shell just ahead of me on the ground. I pointed at it and the policeman leading us nodded okay. My two helpers let me bend down to retrieve it before moving on. This large piece of shrapnel after war's end wound-up as an exhibit on a Red Cross tour in the U S with an identifying card provided by Col Ross Greening, the tour leader. Its card said it was removed from the leg of Lt Joe Reus. Ross knew I'd been shot in the leg earlier, but had somehow erroneously put the wound and the flak together. We then stopped while another policeman emptied the tram of civilian passengers so we could file aboard. They'd have to wait for the next one. Sorry, folks. C'est la guerre.

The tram carried us the fifteen or so miles westward to Oberusal and the main Luftwaffe Interrogation Center there. We got off the tram onto a long train platform and were marched into the camp and lastly into a nearby barracks room.

We were packed in so tightly that there was hardly room to sit on the empty floor much less lay down. Our day of many stops and starts eventually ended with a last bowl of soup. The overcrowding of the room wasn't because we hadn't made proper reservations. We learned later in German air force interrogator Hans Scharff's book about the interrogation center and his part in it, that the crowding was purposely done so that many prisoners wouldn't be able to sleep and would whisper to each other all night. Hidden microphones recorded all that was said. We were also later told that the German air force gathered much more information in this way than through their formal interrogation process. These rooms were also slightly over-heated also to prevent sleep.

During the early morning while it was still dark a very small young gunner next to me told me he was in much pain due to a small but sharp piece of flak in his right hip that was being aggravated by his flying suit every time he moved or was bumped by someone else nearby, and that it was bleeding again. In the half-light I could see the dark blood spot area. I said I could take it out if he wanted me to. He asked me to go ahead.

I worked my way to the door and pounded on it until a hall guard opened it. The roommates wondered what was happening. I pointed to the blood on the man and asked the guard for a messer. He called another guard over, they talked, and finally he nodded okay. He handed me his pocket knife. I opened it, felt its edge, and saw it was sharp enough. With the guards watching I made a small cut in the flying suit and underwear to bare the boy's skin, and using the guard's small battery lamp to light the area, I slit a tiny bit of skin to loosen the piece of metal and ease it out. I gave the small piece of metal to the lad for a souvenir and told the boy to report it to the Germans during interrogation to get it treated and prevent infection.

The guards grinned and nodded as I gave the knife back and thanked the lender. He closed the heavy door while I told the young man to hold his flying suit tightly against the wound and it would soon clot and stop bleeding. I never saw him again. He would have been sent to a different camp. Probably Stalag-Luft 4 for NCO's at Koszalin in Poland near the Baltic Sea. We visited the city of Koszalin in 2006, but the Pow camp, located 25 miles to the southeast of town, had been torn down by the Russians during the late seventies. The area, once open land, is now these many years later overgrown by a small forest. There is a memorial nearby maintained by Polish veterans where memorial services are held annually. We were there to attend such a service during our visit. We learned from former NCO inmates and their wives who there with us that it had been a very brutal regime and that many American died both there and during the forced march evacuation ahead of the oncoming Russian army in 1945.

Later, after we'd heard doors up and down the barracks corridor opening and banging shut for several hours, our door was unlocked and pulled open. A guard NCO, with two riflemen as his escort party, stood in the hall and waved us out, shouting, "Raus, Interrogation" in a loud voice.

Red-headed Irishman, Pete Abell, jumped up and began screaming back, "You dumb bastards, we just got back. What in the hell is going on with you fools? Don't you know what you're doing, you stupid asses." It's a good thing they didn't really speak English. They would probably have shot us all. He went on and on, with a very red face, and put on such a loud shouting uproar, that other guards came to the door to see what was up.

After the guard NCO's conferred, our door was slammed shut and locked. Pete would try about anything. We never were taken over to the interrogation center, but stayed in our barracks room until bread was passed out for breakfast. In the afternoon the guards came back and marched us to another area called the transit section to be outfitted for travel. I don't know whether or not, Pete's antics fooled the guards about the interrogation, or if they were just glad to be rid of the loud bunch of us.

At the transit area we were told that we had to leave all our flying equipment or clothing behind and we would be given other clothes to wear. In my case I was given an RAF blue outfit, because it was the only thing small enough for my five-five and 110 pounds. I was outfitted in blue wool pants, a four-pocket jacket, and a light blue shirt. It even had the RAF " birds" on the tunic shoulders. For shoes I got German GI hob-nailed high-tops. The overcoat was a brown Belgian officer's overcoat fastened by brass buttons with crowns on them. We were all given the soft US GI helmet liner cap. You'll see how widespread these caps were if you look at any American prisoner pictures from Europe.

One of the best things that came through the German Red Cross here, probably from the U S, was a small brown fiber suitcase that contained a pair of one size fits all pajamas, a sewing kit, razor and blades, toothbrush and salt plus soda tooth powder, soap, many other odds and ends, and American cigarettes. Since I didn't smoke, I later found that cigarettes were eagerly sought after by almost all European men and were as useful as money.

Later during the war, prisoners going through the Oberusal transient camp didn't get this useful kit after supplies ran out and German rail transport had almost ceased to exist. Even later, the interrogation center and transient camp were destroyed by a night bombing raid by the RAF and was set up at another location.

The American Red Cross donated to the International Red Cross most of the items that we received here or at the permanent prison camps. Such items were either sent through neutral Sweden or Switzerland to us. We later learned much more about the many things that were sent to us through the diplomatic *back channels*, as I'll mention in later chapters.

After another trip down to the same long train platform on which we'd arrived at Oberusal we were loaded this time into boxcars instead of the tram. The cars were much smaller than American ones. These were what the British call *goods wagons*, and were similar to the cars the Jews were transported in, as you've no doubt seen on TV in films about the *death camps*. Of course, we're told the Jews were packed aboard in far greater numbers and with women and children. We knew nothing of this treatment back then, even though, unknown to us, we had several slave labor camps near our Stalag Luft 1.

Each of our cars had twelve prisoners, plus a minor NCO, such as an *Unteroffizier* (or *buck sergeant*)and three guards. The particular car we traveled in had previously transported coal briquets, and although the floor was covered by a thick mat of straw, we still got as black as apes after a very short time. Because the train conductor had carelessly neglected to attach a restaurant car before departure we traveled for four days with black bread as our only food and with water our only drink.

To be honest, there wasn't really any other way to feed us while enroute . The latrine was the side of the tracks, usually in rain or snow, but could be used only when we were stopped. Our rest stops weren't often or badly needed, because we ate and drank so little. We humans provided the only heating system, but the insulation left much to be desired. If it was a hardship for us, it was shared equally by the German air force GI's with us. We made only the one long trip. I imagine the guard teams made many such trips over many months escorting the large number of airmen who were shot down throughout the war. Some of the prisoner trains made even longer trips than ours, to Stalag Luft 4 in northwest Poland at Koszalin, to Stalag-Luft 6 at Heidekrug in East Prussia, and to other camps all over Germany and the nearby occupied countries.

This duty, however poor for them, was much better than serving on the Russian Front with a Luftwaffe Luft-Land Division (air-landing unit) and fighting as poorly-trained and under equipped infantry trying to stop the Russian horde in the last desperate days of the war. Few would have survived the fighting or the hellish slave labor camps in Siberia. Of the millions of German army and air force prisoners taken to Siberia, only a few thousand came back alive. The SS slogan painted across walls all over German in the last days was, *Sieg oder Sibirien*, which means Victory or Siberia. It was only too true!

One such German army veteran and friend, who beat the odds and made it back from Siberia after five years, was Hans Steffney, a former German infantryman who was shot in the hip at the battle for St Petersburg, Russia, and returned to Berlin. While working in the brown uniform of the *Kraft Korps* (National Motor Corps) at war's end he was taken by the Russians and sent to Siberia. Since his injury made further military service impossible, he worked at a tank factory in the brown uniform. The Russians thought he was a soldier and sent him to a camp in Siberia. After being only one of the few to survive and return to Germany, he later immigrated to the US with his wife who had served with German air force flak units. They later lived for many years in St Petersburg, Florida.

Most of our stops and delays were probably caused by waiting for railway signal lights to change, or to clear the tracks for higher value cargo that was needed at the fighting fronts, or for the war production effort. The delays in many cases caused us to be parked in the same railroad yards that had often been our primary bombing targets, both at night by the RAF and during the day by the 8th and 15th Air Forces. One of the main missions for strategic bombers everywhere is to destroy enemy communications such as this in all forms and shapes. Short-ranged tactical fighter-bombers, such as the P-47 Thunderbolts were used for this purpose, operationally, in the actual combat zones, while we handled the strategic or long range missions. I don't know how many POW trains were hit during the many stopovers, but some probably were hit accidentally. We were the lowest priority freight on the continent and likely used the most out-of-the-way and longer routes.

In the Pacific war, both the US Navy and the Army Air Forces unknowingly caused many deaths when they attacked and sank Japanese ships while they were transporting our own POW's to the home islands of Japan. Many went to their drowning deaths locked in the holds of these sunken ships. The rusting hulks are their only memorials.

We did a lot of sight-seeing through the continuously opened railroad car door. I counted rabbits and farm animals, but didn't spot any deer. One day, while huddled in the open door at the side of a tall older guard, the guard periodically would point and say "*haas*." I though he meant *horse*, even though I never saw one, until it dawned on me that what he was actually saying was *Haas* - the German word for rabbit. Hunting was strictly controlled, but I'm sure the hungry people probably did a good bit of poaching for meat. Goering, of course, had a large hunting estate - and we're told that Rommel did some big game hunting in North Africa between battles.

We saw many large areas of destruction, burned out factories, bridges down, and other signs of our previous raids during our long trek. I particularly remember passing the destroyed town of Anklam, just before reaching our new home at Stalag-Luft 1 near Barth. The industrial area was a blackened ruin. We also got to know each other much better over time with combat, pre-war stories, and general conversations. There was lots of time to kill over the four long days (and nights).

It was then, during the long ride, that a P-51 Mustang pilot told us the tale about his being shot down, captured, and being kept by the German fighter outfit that had claimed him. He entertained us with his story about stealing a BF-109-G, or Gustav, and its aftermath that I have detailed in another chapter. Every one of us had his own "humorous or horror story", too, and wanted someone to tell it to. I guess the saying is true - misery loves company, although a great many stories were very funny. Later in camp we heard so many stories we almost stopped listening, because everyone of us had his own tale. We hadn't arrived by bus.

And then there was poor, old "Alice". Just before and during WWII, the *Popeye* newspaper comic cartoons featured not only "J Wellington Wimpy", who lived on hamburgers, but also "Alice the Goon". This was a strange, funny-looking creature, a female, but not human. One of the German air force guards in our boxcar was a recently drafted *Gefreiter* (Private) with very little training or military aptitude. He spoke English well, and told us he'd been a doctoral candidate at Berlin University. He'd spent his whole previous life in schools and was as far away from being soldierly as can possibly be imagined. He wore very thick glasses, had hips wider than his shoulders, and was very much more overweight than anyone else I ever saw in the *Luftwaffe*. With all this, though, he was highly intelligent, kindly, and well-spoken. Someone named him "Alice" due to his shape and it stuck. Eventually the German NCO and the other two *Gefreiters* (privates) in the car picked up on the name too.

Whenever the train was stopped and one or more prisoners wanted to answer the call of nature beside the tracks, the NCO always ordered "Alice" to climb down and guard them - especially if it was raining or snowing, as it usually was. The prisoners would carefully gather a handful of fine grit, wet or dry, and slip it into Alice's rifle barrel. He never seemed to notice. When they both climbed back into the car, the prisoner gave the high-sign to the NCO, who'd *chew out* Alice and make him empty his gun barrel out the door. He either never learned or paid attention. Later the word *Goon* was used by barracks sentries in all the camps to warn of an approaching German security guard or patrol.

All good things must come to an end, though. Hermann Goering's four-day rail-tour left much to be desired for first time tourists. Our several later trips to the *stalag* area on the Baltic and around Germany have been much improved over that one.

On 1 March 1944 in the early afternoon of a sunny day, at the tiny Barth railroad station, we completed our rail journey and climbed down to the tiny platform, gathered our few possessions, and formed a column for the march westward for two kilometers to the *Stalag*. After being led into a small hall we were given an orientation lecture by a handsome, English-accented, young *Luftwaffe* captain who told us he was a VMI graduate who had been previously educated in England. Our group was next parceled out to our new German compound commander, a Hauptmann von Beck, a WWI fighter-pilot who'd graduated from flying school in 1916. He began his short lecture with "What is expected of young American officers." At the end of his brief talk I rose and said I wished to be the first to ask for his pistol when he surrendered at war end. He stared for a minute and then told me to sit down and shut-up. The good captain never forgot my dig or what I'd asked, because at war's end he made sure I did indeed receive his pistol.

The RAF lads were separated to West Compound, and the Yanks walked a little farther, northwards, to the nearly empty North Compound. It's name wasn't changed to North One until after North Two Compound construction began in the late Summer of 1944. Before we reached our barracks one more delight awaited us - delousing in the shower with kerosene, whether or not we'd picked up any small *critters* during our travels. It's a bit painful until you wash it off with soap, but the German didn't have DDT, sorry about that. We moved into Barracks 7 and had dinner at the mess. *Stalag* is short for *Stamm Lager*, or permanent camp. *Luft* means Air. Our camps were *Luftwaffe* run, but back then by a Germany army colonel.

We were glad finally to once again have a home, even if it was in a prison camp. It was better to once again have your own bunk, your own room and roommates, and a living routine. We were no longer wanderers who didn't know where they were or where they were going. We finally had a place to *hang our hat*. As we got used to our new home, we soon got as busy as a dog with a new bone.

Home is where the heart is, they say - but if the food had been as bad as this at home I'd have slapped my father for marrying such a poor cook. No, I'm joking, it really wasn't all that bad, merely different and somewhat sparse.

Barth is an interesting and very picturesque and scenic small seacoast town on a bay leading in from the Baltic Sea. It's north of Berlin and about thirty miles east of Rostock. Barth, after being founded in 1255 as a Swedish walled town, and it with much of northern Germany, was later captured by the army of Friedrich der Grosse (Frederick the Great) and incorporated into Germany as the province of Pommerania. During our 2005 visit we joined the town's people celebrating their 750th anniversary. We've been back five times and we hope to visit our friends there whenever we can. There are many fine people whom we've met at Barth, but I imagine that all our wartime captors are now deceased, though some of their heirs remain There are several nearby Summer resorts we've visited on the outlying peninsula that separates it from the Baltic Sea near the village of Zingst. Here was also once the Summer *schloss*, or mansion, of *Reichsführer* Heinrich Himmler, but we weren't permitted to make a courtesy call, though near war's end the Russians did so.

The camp, being in the Eastern Zone, was torn down in 1978 during the forty-year Russian occupation. The peninsula on which the camp once stood has now reverted to its original farm land and forested areas. All that remains in the tree sheltered glade is a very large stone marker on which is mounted three bronze plaques, American, British, and Russian, to honor those who were imprisoned there.

In fact, though, we had prisoners from almost all of the countries of Europe, plus many from countries all around the world primarily those who flew with the Royal Air Force of RAF - even Australia and New Zealand.

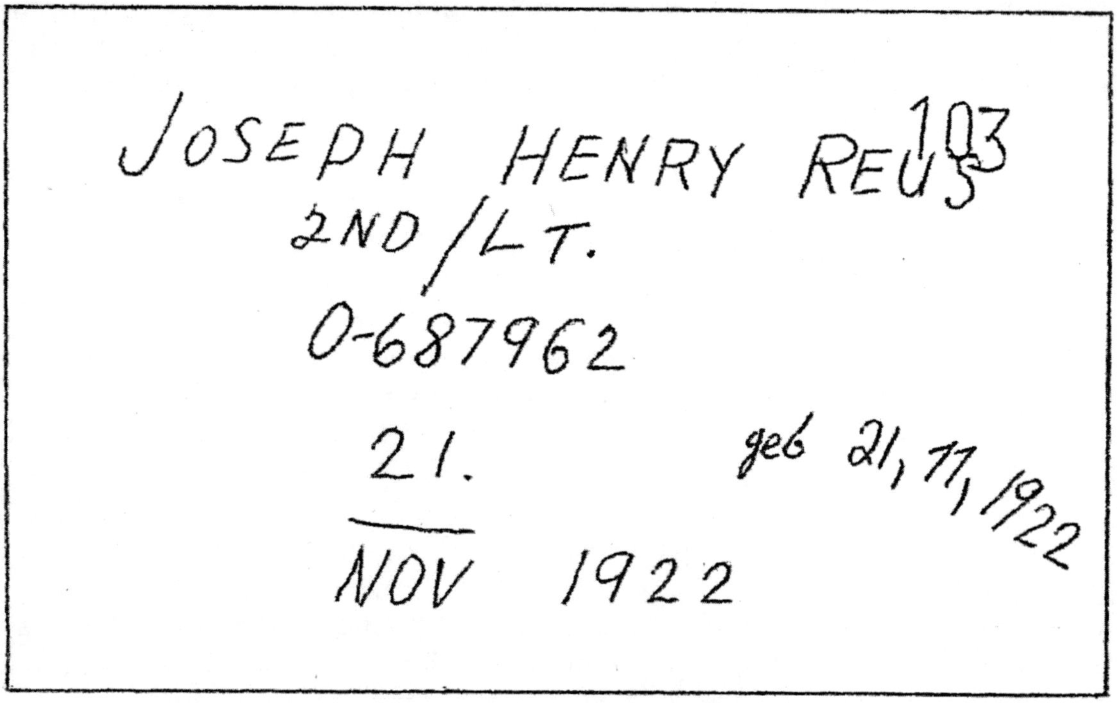

JOSEPH HENRY REUS 103
2ND/LT.
0-687962
21.
NOV 1922
geb 21, 11, 1922

Since I parachuted into Germany without dog tags, probably lost somewhere along the way in my seven-week, three hospital stay in England, the police chief, *Politzei Meister* **Schaller who had captured me, asked me to write the above note you see in facsimile here.**

This note is ***still*** **in the police file in Lauterbach, Germany**

8
Camp Security

Security on Arrival - Identification

As new prisoners arrived, almost weekly, usually by the train-loads in large groups, there was a vital necessity for the Allied leaders to quickly and accurately determine the identity of each individual before they were cleared to mingle with formerly identified POW's. This was to guard against possible German intelligence *plants* or infiltrators. Our camp's primary security officer was Lt Col Loren McCollom.

Not everyone had American *dog tags*, or ID tags, as they were supposed to. I, myself, arrived without them. Presumably I'd lost them somewhere along the route in one of the three hospitals in which I'd spent seven weeks after ditching the B-24 bomber, *Big Joe*, in the English Channel earlier in December 1943. That fact was recently called to my attention by two German historian friends, Horst Jeckel and Robert Keller, who have researched my capture on a snow-clad hillside clearing above the village of Angersbach. As mentioned earlier I was captured by the police chief of Lauterbach, *Politzei Meister* Schaller, and his local pick-up group that was organized to locate and capture downed airmen. The historians sent me a photo copy of the note I'd written for the German police in Lauterbach, the county seat. It had my name, rank, and serial number. Since I didn't have the dog tags they asked to see, or any other ID, I wrote the note when asked. This small slip of paper with my written data is still in the police files of Lauterbach, and is shown in Robert Keller's book *Luftkrieg uber dem Vogelsberg* (Air War Over The Vogelsberg).

Prior to taking part in this mission I'd only been back to our Tibenham airbase for two days. With the squadron commander, Captain Jim Stewart's help, I had gotten a one-mission waiver from our local Flight Surgeon, Major Black, before having to go to Scotland for the mandatory rest leave required after my long hospital stay. In the rush to collect and borrow the various pieces of flying gear needed to replace what I'd lost in the Channel ditching, I didn't even realize that my dog- tags were missing. To have had a new pair made would have taken only a few minutes at the base supply warehouse, but had actually been overlooked. The German police didn't make a big deal over it. I guess I didn't look very "dangerous" with my small size at age twenty-one.

Of course, even with dog tags, there was no guarantee, or even assurance, about identification. The tags could have been taken from a dead airman, a prisoner sent to another camp, or any one of many possibilities. Time was an important issue, because new arrivals could otherwise freely associate and talk with all of the men in the compound. After learning how the long-term RAF officer prisoners from the West Compound handled this situation, several programs evolved to help with this serious issue, but many variations and improvements were gradually Americanized and added. However, these basics formed the foundation of our practices. Some of those British lads were four year inmates, either at this prison camp or had been moved from a previous stalag.

Face Recognition - the first, best, and quickest identification was to be recognized by someone who knew you, even if you weren't remembered by name, and by an American who'd been cleared earlier. This identification could be from home, flying school, squadron, group, or anywhere along the long training route we all covered. No, I never met anyone from my hometown, or schools, but a few did, I'm sure. Sometimes it went the other way, when a new arrival would recognize someone in the compound who'd already been cleared. If the new prisoner had a name or a fact that could be readily verified, it would be a big help and speed-up the process. You can imagine some of all the possible variations.

City or State Clubs - If a new arrival couldn't be face recognized quickly enough, he was greeted by, "Hey, where are you from?" When the new man said what town, state, or place he called home, he was "invited" to join that "club" and meet fellows from there. Over a cup of *ersatz* (substitute) coffee, the new "friends" would chat about home, high schools, ball teams, and on-and-on, but nothing military.

The "friends" would very quickly find flaws, if there were any. Unlike the movies and story books that use various *gimmicks*, no *plant* could pass the test for more than a few minutes. German intelligence, or any other, couldn't possibly know or cover all the needed answers, because they aren't written down anywhere and couldn't be studied by a plant, even if an agent could possible remember all of the myriad possibilities. For a new man to hang back and not join in would cause suspicion and also might be a giveaway, or engender doubt.

You also gave the new man a few inconspicuous false items to see if he'd "bite". *Mission Impossible,* is just that. We all took part as we fitted-in when the need or opportunity arose. The mountain of simple questions and data we could create was virtually insurmountable. Most new arrivals probably thought we were being very hospitable and didn't realize they were undergoing a very fast, thorough, and special interrogation. I had arrived with the other three officers I'd flown with and came to the same compound at the same time, so our acceptance was easy. I doubt if the Germans would ever have attempted a four member *insertion.*

In one specific case what happened with the men in our room 7, barracks 10, was a good example. A young man arrived who told us he was from Louisville. He flew P-47 Thunderbolt fighters with a newly arrived, but unknown, squadron with the 15th Air Force in Italy. He had a German accent, and was named Steiner. Accent wasn't really a very good or reliable guide, though. We had other German-Americans in the compound also with large or small accents. Nobody recognized this man's squadron number or him, but he knew quite a bit about Louisville and his high school. He was assigned to our room, because we had a Lt "Mac" McConnell, a B-17 bombardier who'd grown up in Louisville. Mac, who played guitar with the compound's orchestra, often practiced his country music in our room. One morning, while Mac was strumming his guitar, he boasted that he had played early morning country music on Louisville radio station WLAF. Steiner chimed in from the bunk above where he'd been placed deliberately, with a German accent, and said, "In Louisville they call that WLAP - We Let Anybody Play." Since that was a local joke, which only someone from that area could possibly know, it not only brought the house down with laughter, but it also identified Steiner.

Lt Steiner is the P-47 fighter-bomber pilot I wrote about in the *Chivalry* chapter, who crash-landed in Italy with his tall and skinny crew chief jammed into the single seat under him. The crew chief had always wanted to fly a combat mission, especially a dive-bombing mission. For a *lark* they tried it out. He was later fully cleared when his squadron commander arrived after being shot down and was himself recognized by other fighter pilots. The crew chief would have been sent to an NCO camp in a different area.

Un-cleared men were *taken for a walk* when the barracks news-rep came around each evening just after dinner and before the barracks doors were locked by the German security patrol. The news rep gave an un-classified briefing condensed from the local papers - and an oral-only classified section of the daily BBC radio news excerpts. The BBC gave *dictation speed* news which was received by our hidden radios, as it was by all the clandestine or underground groups all over Europe and Asia. We had radio receivers in both West and North One Compounds. I don't know if North Two or North Three had them, since these compounds were built very late in the war. Radios were assembled by our communications people and hobbyists from individual parts bought for us by helpful German NCOs and officers. I have no direct knowledge of this, since I was a consumer, not a provider, of news. Colonel Hubert L Zemke, our last Senior Allied Officer, gave all the radio details in his book, *Zemke's Stalag*. We had all sorts of skills available in many fields among the American, British, and other Allied detainees - and we had many of the brightest people each of these countries could provide.

Hearsay had it that a lieutenant in Barracks 4 never was actually cleared. He said he was from a very small town in the Midwest that no one had a line on. No one knew him or his area. He gave out very little that could be verified. He had a squadron and bomber group number, but said he was a replacement who went down on his first mission and knew few people. He was a loner who didn't make friends. One day the Germans took him away. Ostensibly, he was sent to another camp. This alone would have confirmed our suspicions. I know no more about this story than that or if it was true.

Goon Guards - We took our turn as sentries, or as we called them, *Goon Guards,* at the center hall doors at either end of each barracks to warn anyone doing confidential work to cover up during the approach of anyone in a Luftwaffe uniform or civilian clothes. We all did it as a routine and took turns doing our tour. If we'd done it only when something was being hidden, the Germans would have eventually caught on and put two and two together and showed up when we mounted guard.

No, these people were far from stupid. In fact they were not only very sharp, but they had learned over many wars how to securely keep prisoners. We learned many things from the Germans and their activities in an attempt to evade their security efforts. I'm sure we all learned from each other.

One day when I was on guard at the south door a burly guard approached. I sang out, "Goon Up." The guard scowled and asked, "Vat iss dis Goon?" I realized that he spoke some English, so I smiled and said, "Oh, it means a minor official, like a deputy mayor or such." He gave me a big grin, nodded, came in, and walked to the barrack's captain's room to conduct whatever business he came for.

Covers of all sorts were created to conceal our activities, or create opportunities. As an example, the Art Committee, for which I was the mechanical drawing, or drafting, instructor, was created not only to teach various forms of art and drawing, but it was also the *Map & Document* section of the Escape Committee to provide map tracing copies, gate passes, and the like. By now, you're probably aware that the RAF had cloth maps sewn into their uniforms for just such occurrences. There were many such things wanted from us. We provided what was needed. Some worked, some didn't. I have a few more examples in the Escape chapter.

Many outdoor construction projects were undertaken such as re-building our softball diamonds or volleyball courts merely to conceal new dirt from a tunneling job. The German security police probably saw through our many schemes, because they would have seen the same efforts in previous wars and at many other prison camps, whether they were military or civilian. What pretexts and pretenses we thought were new were probably as old as time. Over the length of our stay we found out how sharp their security people actually were.

Whenever some sort of noisy construction was needed that we didn't want the German tower guards or foot patrols to hear and wonder about, we booked a practice schedule for the compound's Men's Choir to furnish sound cover. Their version of the Welsh fighting song. *Men Of Harlech.* would rattle the window panes and would cover almost anything up to an air raid or natural disaster. Their leader was a song writer and musician who also led the compound's orchestra, *The Round The Benders*. This man, Harry Korger, was the lead navigator on the first bombing raid over the oil refineries at Ploesti. Years later he became the chief navigator for the

Strategic Air Command at its headquarters near Omaha.

German Camp Security. The German security forces arrayed against us were at many levels from the barely trained to the very well trained and with very intelligent NCOs and officers. The camp's German intelligence chief, Major von Mueller, even created a museum with models of various escape attempts. I'm told that these models were also used as training aids for the security patrol personnel and tunnel-locating *ferrets*. I saw the tunnel escape models after the war ended when the German commander turned the camp over to American control. They reminded me of the ant farms with glass fronts we saw as children. They also reminded me of my first trip to the *klink*, after our own tunnel dig was discovered.

The lowest level of training and experience was among the exterior walking guards who patrolled the outside of the fence lines, both during the day and after dark, in freezing rain and snowstorms. They were potentially the most dangerous of all, because they were easily startled and might fire their rifle if frightened or even if suspicious. These were much older men called-up from every walk of life in the latter days of the war. They had little if any real training unless they'd served in WWI or had other previous military service. Such men were used inside the compound only during roll-call or, *Apell* in daylight hours or other times when they could be led or controlled by regular Luftwaffe NCOs.

Tower guards were also usually older men from lower NCO ranks. When possible there were guard towers on each compound's corner, or the towers were located at each fence end where two compounds met. The towers were equipped with single or twin light aircraft machine guns fed from small ammunition drums called *snail drums*. For night use the guard tower had a strong searchlight which was rotated randomly. Naturally the lights were turned off during air raid alerts. This gave us an opening to prowl about, cut wire from fences, and steal boards from buildings. You could be shot or arrested by a patrol, bitten by a dog, or moved to an open bunk in solitary, but most of us tried the *night life* at least one or more times.

There were random security patrols during the day or night by German air force security police personnel. Some of these patrols were accompanied by dogs, usually non-barking German shepherds. Most of the dogs did not seem to be very well trained and could be befriended when they were allowed to run loose. If we saw a free dog after dark we would entice it near and feed it meat scraps which we hoarded for this purpose, which the dog probably didn't usually get, but really wanted. As soon as he learned the smell of your window and you, he'd come back whenever he could. When they stuck a nose into the open shutter we'd pet them with as many hands as possible - so they would "know" as many people as possible.

Our windows were covered at night by big wooden shutters, or *fensterladen*, held shut with a wooden peg through a hasp, but they were easily opened with a knife through the crack from our side. Some men went out after dark after they'd gotten to know the dogs, for such things as stealing barbed-wire from the fences. I did several times, normally during air raids blackouts, but sometimes in 'blind spots,' or areas not seen very well by tower guards. If you remained still the guard usually wouldn't see you. You would be okay if the roving light didn't stop or return for a second look. The stolen wire, after the barbs were removed, had many uses.

Ferrets were the small Luftwaffe security men who crawled under the barracks at night looking for tunnels. They probed in the sand with long metal rods looking for tunnel shafts. In one case, men in another barracks, perhaps guarding a tunnel, or just for devilment, wedged their floorboards apart slightly and waited for the sound of a ferret probing underneath. When they heard one, they pored down a bucket of urine. The guard was so"pissed-off" he emptied his Luger pistol up through the floorboards. Fortunately no one was hit. The room-mates had to carve wooden plugs later to fill the holes and cut down the Winter-time draft.

Although I wasn't aware of it until the war was almost over, Col Ross Greening told me that the Germans had a seismographic team, with small, plate-sized units near our fences, to locate tunnel digging. I'm sure none of us knew that when we were moving so much sand. It was good, if dangerous exercise due to the potential for cave-ins in the soft sand. Of course, we did use shoring, made from wood scavenged from the barracks for support. As far as I know, no one was trapped in a collapsed tunnel, but I'm sure there were quite a few "frights."

The gradual *combing out* of all younger, able-bodied men for East Front service with the Luft-Land divisions (air-landing) never stopped as the war went downhill. Later, after the war ended, Col Greening managed to ship Major von Mueller's tunnel models back to the US for his national tour in support of the Red Cross, and for our government's intelligence uses.

After the attempted bomb plot, in July 1944, by Oberst Klaus Shenk, Graf von Stauffenberg, against Hitler at his East Front headquarters, the Wolf Schantze, or *Wolf's Lair*, in East Prussia, many changes were made throughout the German government, and they even reached our small outpost. The only change that affected us slightly as prisoners was Hitler's order to Himmler giving his *SS-Sicherheitsdienst* (Security Service) control of all military prisoner camps, in addition to the political prisoner camps he already controlled. Those were called *Konzentrationlagers*, or KZ's. This Hitler order, and all other Berlin orders was posted on the bulletin boards of each barracks and produced the usual groans.

Previously, military prison camps had been run by the military services. In our case it was the Luftwaffe, under the command of its commander-in-chief, Hermann Goering. This was a huge gain for air force prisoners due to the almost inherent chivalry that seems to follow all flyers, and the fact that Goering's air force and political position gave us greater access to the food and other things we needed.

When our group of prisoners had arrived in March 1944, though, the German commander was an army colonel, Oberst Scherer, while all of the other officers and NCOs were Luftwaffe men. Toward the end of the war, Scherer was transferred and the new Air Force commander was Oberst Warnstedt. After Hitler's order to Himmler to take over all the camps, we did see a small number of *Heer*, or Army NCOs, but to the best of my knowledge, there were never any SS men assigned to our stalag.

One German Army NCO, newly arrived, had formerly been an American immigrant and had worked as a butcher in Newark, NJ, before the war started. His return to the resurgent Germany turned out to be a poor choice. He was known to one of our men, a B-17 co-pilot, who had lived on the same block with his butcher shop. Naturally we called him "Henry, the butcher." I have more information about him and his antics in the *Liberation* chapter.

West side of North 1 Compound, showing warning wire to the left.

Partial view of barracks in the foreground is Barracks #9 where the senior officers lived.

Middle barracks is #10 where I lived about half-way up this west side.

End barracks is #11, a former warehouse converted to Barracks 11, later called the Jewish Barracks.

9
Cuisine

We often jokingly described our prison camp food as our *cuisine*, because food was a very important element in the life of a *Kriegsgefangener*, or war prisoner. Prison camp cuisine, or as some called it -*Barf On The Baltic,* really wasn't bad, and we were probably fed as well as possible by the German Air Force under the existing circumstances - and with the Red Cross parcels, better than that which was received by many German civilians.

Most of our food came from two primary sources, the limited number of German issued ration items and the Red Cross eleven-pound food parcels. Without the first we would not have had the needed "bulk,"such as potatoes, kriegsbrot (black war bread), cabbage, rutabagas, margarine, prunes, strawberry jam, pea flour, turnips, etc, and without the second we would not have had the needed calories to maintain health.

The Red Cross parcels were very well prepared, and although we did not always get one per week as was no doubt planned, over time their delivery overcame many shipping problems. Our strategic bombers' destruction of the German transportation system, primarily the railroads, was the single largest limiting factor. The fact that these highly valued food parcels made it safely across the German nation, to us through a hungry population, always amazed me.

I believe a few of the more infrequent or seldom seen food items were bought with the pooled *kriegsmarks*, of the imprisoned officers' pay. Officer POWs were paid so much per month, depending on rank. This money was not received individually, but by our paymaster, Captain Dixie Lee, under the terms of the Geneva Convention from the German government through the Luftwaffe. Among these probably bought items were such things as the two small rations of beer and the infrequent delivery of such foods as sauerkraut, and other occasional foods when available from our very small town. More about Dixie Lee later.

The Red Cross food parcels were delivered regularly during our earlier months of confinement before German transportation was gradually crippled by our strategic bombing campaigns against German road and rail traffic. During the final months of the war rations became fewer and arrived farther apart. Sometimes we had to split parcels when there weren't enough to go around.

At Christmas 1944 there was even a special and separate Red Cross parcel, both for the US and British Pows. Ours contained canned turkey and plum pudding. I believe the British ones came with canned roast beef and other English favorites. Because there were too few of the British ones, and they liked ours better, we helped the *lads* out by sharing. This seems to be an American trait in poorer times. Probably *Lease-Lend*.

In general I'm sure that we were much better supplied than most of the German population during almost any period from mid-1943 through the end of the war. In German cities many people actually died from starvation as they had done in the aftermath of WWI and I believe under these circumstances county people living on the land and growing some of their own food did far better then city-dwellers.

The German ration items were listed as so many grams per person, per day. Included was bread, or, as we knew it, *Kriegsbrot*, or black *war bread*. It was made from any grain other than wheat, and it was made from whatever grain that was then available. The bread had a very dark color and sour taste, much like a very dark pumpernickel. Despite that, it had far more food value than our white bread, which we wouldn't have survived on. At first we brought it back from the mess-hall and toasted it on the German tile stoves in our rooms, but many of us later grew accustomed to its taste untoasted. Without this bread, usually made at a German Army bakery unit nearby with the help of Polish conscripts and later American volunteers, we would have been in far worse shape later in the war when other food sources began to dry-up and become almost non-existent. This bread was also a mainstay for German civilians who would have died without it. Many of the people, especially young children, had bloated stomachs due to a diet relying heavily on potatoes and little else.

With us, white bread, made from wheat, was only available to those patients for whom it was prescribed by German doctors for ulcer treatment, recovering war wounded, and serious illnesses . In our case, it was found that flyers have a very high frequency of ulcers. It was also given to other, non-military, patients for whom a lighter diet was needed during recuperation at hospitals. Americans aviator prisoners got it only when it was prescribed by the doctors at the Luftwaffe Orthopedic Hospital at Neubrandenburg, again, either with ulcers, or when recuperating from injuries or wounds.

That large hospital was then among the world's leaders in orthopedic surgery during this time period. It was located about halfway between our Baltic Coast camp and Berlin. Ulcer patients and the severely wounded from our camp were sent there for treatment earlier during the war when transportation was still available. Several prisoners, mostly RAF men from West Compound, were still alive only because of them. I personally never saw any of our people receive white bread.

One amusing, though erroneous, rumor arose about German wartime bread production. In order to save flour of any sort, flour was not used to dust the board where the dough was rolled and cut. Instead, ground wheat chaff was used to cover the bread-board. This left the dough, and later the loaf, with a crust looking like sawdust. The rumor spread, even to the U S, that the Germans had found out how to make bread from wood. There was even a Disney cartoon movie made during World War II which showed trees being cut and processed into saw dust and then baked into loaves of bread. We wondered if some of the dim-wits in the Pentagon in Washington actually believed that digesting cellulose was possible.

The primary breakfast cereal was boiled barley that was cooked in large double boilers in our mess hall kitchen. Sometimes, on Sundays, or special occasions, ground chocolate from the *D* bar, provided in the Red Cross parcel, was added to make *chocolate barley*. It was a real treat. Another well-liked Sunday morning favorite was creamed Spam from the Red Cross parcel on black war bread. Believe me, it was better than it sounds now. But like the old joke, you had to be there.

Another standby was the eight, *count 'em*, prunes made available each morning to supply us with vitamins and minerals. These prune plums have been grown from ancient times in vast number in Germany and all over Europe. They are dried and stored for winter use by almost everyone for taste and health.

Periodically we received a ration of red "strawberry jam," so-called, even though it contained no strawberries. It arrived in large wooden barrels by wagon. Instead of strawberries it was made from the sugar beet residue after beet sugar was extracted. Somehow the makers converted what would have been a cattle or hog feed base product into a tasty and attractive food. It was colored red and flavored, too. We even joked that the Germans threw sand into the mix to make it taste like real strawberries.

One of the occasional offerings from the Germans was pea flour. We not only made the thick pea soup most people are familiar with, though we didn't have ham for flavoring, but someone came up with the idea of making pea flour pancakes. To make them rise, like normal flour pancakes, we had to have a room mate go on "sick call" to the medical clinic in West Compound the day before to pretend he had an upset stomach. There he'd be given the usual sodium bicarbonate pills that when ground became our baking soda. *Voila!* Afterwards the word spread and there was a "run" on pills. The British doctor at the medical clinic often asked, after he was told about an upset stomach, "Oh, pancakes tomorrow?" In any case, the pancakes were very colorful and were quite filling and tasty, even if a little strange. We also hand-ground crackers to make flour for this and other purposes.

Typical Contents of 11 lb. Red Cross Food Parcel. Parcels varied in contents over time, but main items continued even if from different labels and suppliers. Four parcels were in each larger carton for travel.

Dinner was usually some combination of potatoes, cabbage, rutabagas, black bread, and whatever meat was available from the food parcel. Canned corn beef was usually cooked with potatoes and cabbage as an Irish dinner. The Spam provided was often creamed with the powdered milk, Klim, and served over *Kriegsbrot* on Sunday mornings. Sometimes Spam was lightly fried and served with potatoes and cabbage.

If you didn't like boiled cabbage you were in the wrong mess. The wiser men hoarded non-perishable foods, such as these meat cans, for hard times. You got thinner as time went by, but as you gained experience as a prisoner you never really went hungry very often. All of the foods that the Germans supplied were those that lasted well without refrigeration. We had none. As time went by we became wiser and didn't *bash* the food. Bashing the rations was an RAF term that meant stuffing down all your food without concern for tomorrow. Only fools, or new prisoners, did.

From our first arrival at the stalag we had eaten our two meals, breakfast and the evening meal in the mess hall, until many months later when one night the mess hall suddenly burned completely to the ground, The Germans didn't arrive soon enough with their very limited fire equipment - and all we had to use were buckets that got us wetter from the bucket-brigade than the building did.

I don't actually know if the burning down of our mess building was arson or not. But here are the few facts as I remember them: The fire occurred shortly after a large group of young lieutenants had a *beef* with the senior officers about who decided how much of what food was taken from Red Cross food parcels for use by the mess hall kitchen staff in meal preparation. The senior officers told the hotheads that we were military people who didn't live in a democracy, and that they, the seniors, would make any and all decisions. Questions, any one?

On the day before the fire, our small *exchange store* and the band instruments room were all emptied from their storage rooms at the mess hall the night before, with nothing lost to the fire. The exchange store was a small anteroom just inside and to the right of the main door of the mess, before you passed through another door into the mess. It was a place where you could swap a food can you didn't want for another on hand that you liked better. The exchange fee was two cigarettes per item.

The musical instrument storage room was another anteroom, the mirror opposite of the small post exchange store and was across the mess entrance from it. Somehow, all the musicians had decided to practice with their instruments in their individual barracks the night before the fire. Did they know that the mess hall would burn down that evening? Do you believe in miracles? You decide.

After our mess hall burned down, to cook the large volumes needed by the compound's population, items like boiled barley, potatoes, or cabbage, and similar things, the Germans provided the compound with several large army field kitchens, with shed roofs over them for weather protection, to fill the cooking gap. They even taught our volunteer cooks how to use them properly (even when it snowed). The other small volume foods were then cooked in each barracks room over the tile stoves which now did both mess and heating duty. The other compounds, unlike ours, had never had mess halls and had always cooked in their barracks rooms.

Here I might mention that like most other prisoner barracks rooms, we had modified our room's tile stove for convenience long before the need arose for cooking meals. The plenum chamber, normally used to provide room heating, was converted to also serve as an oven. The firebox served as the barbeque and roaster/toaster in addition to a quick heating source for water for *ersatz* coffee and my herb tea. The top, always warm, was the slow cooker.

It was surprising how quickly new ideas popped up and if suitable were adapted by many if not almost all. Many good ideas were adopted from the RAF, who had learned a great deal about camping out in a barracks room, with no mess hall, over their many years of confinement.

The RAF set aside common rooms for cooking and other such uses in their barracks for meals and their *char*, which is what they called hot tea.

During the Christmas and New Years "Open House" in 1944, when the whole camp was placed on parole and we were free to roam around and visit the other compounds, we were able to visit the RAF compound along with the other two compounds. We could see "how the other classes lived". Vegetables provided included potatoes (*Kartoffeln*), cabbage (*Kraut*), and rutabaga (*Kohlrube*). Because we lived in a mostly rural area, these foods were readily available. They stored well and were also available all year. The vegetables and bread were our biggest volume foods.

In the Fall after the harvest, these foods were stored in the fields in which they were grown. Farmers plowed large, long, *V* shaped ditches, filled them with potatoes, cabbage, or rutabagas, etc, then covered them with a yard of dirt to prevent freezing. The farmer, who knew where each items was buried, then could dig into the mound, uncover the amount and type of vegetable needed, and re-cover the mound for later use. In America, the farmers mostly use potato barns where the crop can and often does, as in Maine or other northern states, occasionally freeze and be lost. I don't know of any other storage for cabbage, etc.

This ancient storage method was also a good food source for escaping prisoners or the German needy. Prisoners with an approved escape plan were well briefed on this valuable food supply source along their planned escape route. I'm sure the prisoners and guards being evacuated from camps in Poland and East Prussia knew about this and made good use of this food, during their long, freezing march back to Germany, even if the food was cold and raw. When you're down to bare survival, anything and everything tastes good.

Many other foods became available from time to time, such as the canned beef Hitler had bought from Argentina long before the war. To buy more meat with the money he could afford, he skimped on the quality of the can lining. It was a poor and unwise bet. Many cans rusted through during long warehouse storage and rotted. We were offered it, but there were few takers - even from the hungry. It was safe to eat if cooked well, but few of us could stand the smell.

Another smelly item was the foil-wrapped, triangular-shaped, soft cheeses, called *Greif* (Griffon or gargoyle). You may have noticed that Feldmarschall Erwin Rommel used the same name on his small SdKfz-250, armored personnel carrier, in the North African war. Like *limburger*, few Americans ate this cheese. Those who did, probably had no "close" friends. We buried some under the barracks in the dirt to get rid of it. We did not refuse anything offered.

Sauerkraut was a sometime favorite. It, like the B*rot*, and many other bulk foods was brought by horse and wagon from the town of Barth. The difference was that we could smell the kraut almost as soon as we could see the wagon. No one makes it better. We loved it then and still do when we get back to Germany. In WWI the Yanks even called Germans *Krauts*, (cabbages), because this was an important source of their diet back then - and during this war.

Another "sometime" food came by wagon without a horse pulling it. The horse rode on top of the wagon, while twelve of us on each side of the long wagon tongue pulled it. Grazing horses were staked out on the airfield with chains to eat circles of grass. When horses were accidentally killed, or too badly injured to be saved, by low-altitude strafing attacks from American P-51's or British Mosquito fighter bombers, the camp got what was left of the animal, in "whole or part".

American P-51 Mustang fighters that escorted the heavy bombers in the daylight raids later in the war, came in waves. When each wave was relieved by the next wave, the first wave planes were released. Many dove down, singly, to strafe targets of opportunity such as airfields with parked planes or vehicles. Even though Germany was the originator of the *Blitzkrieg*, or lightning war, using fast-moving armored vehicles as spearheads, most units relied heavily on horse-drawn vehicles. This was especially true at the beginning of WWII when about eighty percent of German army vehicles and artillery were pulled by an unbelievably large number of horses. They, too, ate all the killed animals. In bad times such as at the siege of St Petersburg, they killed all the horses and dogs for food.

We had to go and get the horse carcass by pulling our big wagon. The horse was then carefully skinned and butchered after it was brought back to the camp. To us it was German prime rib. Each compound got its turn at the banquet table. I doubt if anyone turned their nose up or complained at this bounty. Horsemeat is still appreciated all over Europe and other parts of the world, if not here. Because of the American love of horses, few of them are now killed for meat. The ones that are used for food are consumed quietly - mostly in farm areas away from the cities and in northern areas whose resident's ancestors came from northern Europe. In most other parts of the world, horsemeat is a readily accepted part of the diet if and when available.

An unusual German ration item was the large amount of sugar given to us during a period when little food was available. We lived in an area that produced large volumes of sugar beets. When the limited rail traffic could not ship food in, or ship processed sugar out, the German *kommandant*, Oberst Scherer, requisitioned some of this sugar to supplement our poor diet. We received boxes of it, and because we could not eat it as is was, we had to learn candy making.

That sounds like Marie Antoinette. When told that the French people during their revolution against the monarchy didn't have bread to eat, she's supposed to have said, "Let them eat cake." In our case, it was candy. I learned to caramelize sugar and to make other varieties. We also ate all our failures.

My favorite candy was a milk chocolate made from Klim, the powdered milk from the Red Cross parcel. The finely ground dark chocolate from the "D" bar with lots of sugar was combined with powdered milk. The "D" bar was a non-melting chocolate bar that soldiers were issued in tropic climates. The Red Cross was wise enough to include it in the parcel for us. We didn't live in an area that was hot, but then, neither was much of our everyday food. These things, like certain canned goods were the things we squirreled away for darker and hungrier days as the war wound down and the food supply became more "iffy".

As was previously mentioned about the more rare distributions, we were twice given a small ration of the local beer brought out by wagons in large wooden barrels. I don't know if it was a gift or was bought with the *kriegsmarks,* (war money that could only be used through the camp system), we regularly received from the Luftwaffe to our paymaster. Though officer prisoners were paid different amounts, based upon rank, all present got the same share of whatever supplemental food or other items that were bought, such as razor blades. NCOs do not receive such a bounty, but received the same share of whatever was bought as a colonel did.

Other rare foods, such as donations from foreign countries, while fairly rare, did arrive now and then. In one foreign donation, I recall, was a shipment of oval fish cans with a fish stamped on top that came from Argentina. The cans were similar to the ones that hold large brisling sardines. When we opened them, however, we found apple butter. It was great - and we were very grateful, even if the can didn't contain fish. No doubt I've forgotten as many other special items as I remember.

Here's one shipment we didn't get. In one funny case, an RAF Mosquito fighter-bomber attacked a train containing Red Cross parcels just after it arrived at the tiny Barth railroad station. The train was set afire by the aircraft cannons and was almost totally engulfed in flames. All the food was lost. The only car cut loose in time to save it by the *kriegies* unloading crew was the last car - it was full of toilet paper.

During the warm, golden Summer days of '44, when we had plenty of potatoes, we found that we could make french fries even though the potatoes had already been lightly boiled. This gave us a late night snack, since that far north in the middle of Summer, the Sun stayed above the horizon until well after eleven.

Cooking oil was saved from canned Spam and any other source, such as the substitute margarine. Oil had many uses. One use was for lamps after all the water was cooked out. It burned very brightly in our *kriegie* lamps. The wick was made from pajama string, though other cords would work almost as well.

An empty six ounce jelly can, similar in size and shape to our present-day tuna can, from the parcel when emptied and cleaned, had its lid bent vertically. Halfway up the lid was bent back horizontally over the can. Through a small center hole in the bent top lid, made by a nail extracted from the building, the wick ran down into the oil. When the Germans turned the electrical power off at ten, anyone wishing to read, play cards or chess, or do model work, lit his small lamp. We were supposed to put all lights out during air raids, but the heavy wooden window shutters didn't let much light out anyway.

Yep, we had ice cream too, but only in the Winter. With no refrigeration or even a compressor to create cold, we couldn't make anything cool in the Summer. We had to wait until Winter provided icicles, so that we could use salt and ice to make the needed cold brine. All you needed was a bucket for the ice and a Klim can, with a replaceable top, for the mix. Sugar, powdered milk, "D" bar chocolate, or any of the various jellies that came in the food parcel made up the mix. It gave us many varieties, some rather wild. Sorry, no pistachio or rum.

With a nail driven in near the rim at the top of the closed Klim can, we sat and swirled the can while we read a book held by the other hand. Periodically we removed the can top and stirred the mix as it hardened. It was ice cream and overcoats.

German margarine was made from denatured coal tar. It had no food value but it didn't bother your stomach like mineral oil, tho' it helped make the bread go down - and it was also a fuel for your lamp. It was cream colored, like butter.

To add flavor and salt to this margarine, we melted oil from Spam and heated it with the margarine, then let it cool to re-harden. Another use for it was as the base for "make-up", for stage plays. Added to it were ground water-color pigments, or any solid dry colors from the art classes, to get the desired shade.

The eleven pound Red Cross parcel was a *godsend* that provided food and many other things that the German couldn't or wouldn't provide. There were two 12 ounce cans of meat, usually Spam and corned beef. A six ounce can of pate (most of us tried to trade away the liver type), a 16 ounce can of powdered milk called Klim, an 8 ounce box of sugar cubes, a 6 ounce can of jelly, a 12 ounce chocolate *D* bar, five packs of 20 American cigarettes each, soap, salt & pepper, plus packets of coffee. The RAF Red Cross parcels had tea, but it turned our teeth black. I've probably forgotten an item or two. The empty cans and even the cardboard box were put to use in many ways, such as shelf making, models-making, etc.

The 100 cigarettes were used as our currency. Each cigarette equated to one mark. They were used in trade among ourselves, and with the German guards. We weren't allowed to trade individually, but only through each compound's designated, German-speaking, American *trader*. In this way, there was no inflation, as there would have been if we'd bid against each other. It was also sanctioned by the German command. I didn't smoke, so my "riches" piled-up, even if there were very few things to buy. I heard that bicycles went for six hundred marks, but there were none "on offer".

On warm, dry days some enterprising individuals from barracks three even brought out tables from the barracks, covered them with blankets that had numbered squares printed on paper, to form a "casino" with a large numbered wheel. I don't know if they got rich. If they had done so, there was nowhere to spend their "take". The guards hungrily stared at all the American cigarettes.

For quite a while the Germans took the pepper out of the Red Cross parcel, by having our ration gang pick out the pepper and turn in a packet for each parcel opened. When the packets arrived, with mixed salt and pepper, we lost both. The ration group took out the food items needed for the mess hall meals as I noted before.

What we learned from the Germans, and hadn't known before, was that pepper can be used to prevent tracking dogs from following a scent and locating escapees. A flake or two in a footprint causes the dog to lose his ability to track for up to a day. It's not harmful to the dog. He's not even aware of it. In the US, escaping cons are reputed to have used kerosene for the same purpose, to throw the dogs off, but a sharp-nosed guard, supposedly, could sometimes smell it.

Eventually, our senior officers worked out a parole agreement with the German command about the pepper, and everybody was happy. The Germans wouldn't take it, and we wouldn't use it during escapes.

I forgot to mention turnips. While in camp I made two vows for after the war: The first was to never eat another turnip or rutabaga. The second was never again to leave something uneaten on my plate. My waistline shows it. Unfortunately, I probably won't ever get to heaven, since I've broken both vows.

Though this isn't connected to food, we were also able to work out a parole about the wood-fiber filled mattresses we slept on in our wood-slat bunks. Mattresses were filled with a coiled wood shaving called *excelsior*. It was comfortable, warm, and could be "fluffed up" when matted-down after long use. The mattress "bag" was woven from a longer wood fiber, like a thick brown thread, into a rectangular shape vaguely similar to our twin-size mattresses.

The parole arrived at by our senior officers with the German staff was that we wouldn't hide contraband in mattresses. The Germans agreed that mattresses henceforth would be free from any future searches by their security police patrols. It was a good deal, since I'd sewed overcoat buttons on to my mattress, with button-holes in the sheet and blanket, so I could button down the overcoat over me during the coldest part of the Baltic winter. Necessity is the mother of invention - or something.

Towards the war's end we received small amounts of dehydrated applesauce. Dried, it looked like sawdust, but when heated with water and sugar it was excellent. I even made a final pie from it in a baking dish made by the tinsmith group. This group rolled out the tin from used cans and made all sorts of useful items. Only a few of which are frying or baking tins. The Germans had earlier wanted the empty cans turned-in for their scrap drives, but later allowed us to keep what we needed.

Herb tea was another one-time handout. I was given a large box of it to ration out to the men in our barracks. It looked like dried grass-cuttings, leaves, and pieces of weeds with a few wizened berries of odd colors thrown in. Strangely enough, I couldn't even give it away, not even to the RAF enlisted men who normally drank tea and lived in the small anterooms at each end of the building. Being a tea drinker, this didn't disappoint me. I stuffed the large box under my lower bunk and had enough tea to last until the end of the war. My Scottish-born stepmother would have loved it. As children we were given herb tea when we had a bad cold or fever, and I never lost the taste.

Side dishes, like eating Senior Allied Officer Col "Hub" Zemke's cat, and the guard dog that was trapped, killed, and eaten as described by Bob Buecker and his roommates in North Three compound, were rare events, and were done more from devilment, than need.

Finally, I was probably one of the few prisoners who ever ate seafood, other than fish, while a prisoner. Our camp, unfortunately, never received fish during my stay, even though our town of Barth was a fishing port.

I had a side dish during the summer of 1944 when I was able to collect black mussels from the tall water reeds that bordered the Boddensee. It was the body of water that lay to the west, north and east side of the peninsula on which our camp was located, and where we swam while on parole hikes. Others in the group, except for the German Luftwaffe NCO guard / escort who went with us, didn't recognize them. They had never eaten them. Each time I could, I carried them back to camp and steamed them for dinner. I offered them to roommates, but as with the herb tea, I had no takers.

The gradual destruction of the German rail systems made the delivery of our food parcels at first difficult, then almost impossible. To bridge this for us, a very large gap, a little known operation was put into place. It involved the cooperation between two deadly enemies - the US and Germany. It was set up *under the table*, or as a *back-channel* operation, in diplomatic parlance.

It began with the donation of a fairly large group of *18-wheelers*, standard American Army 6 x 6 trucks, to the International Red Cross in Switzerland. These vehicles were loaded with the maximum load possible of the forty-four pound cartons that contained four Red Cross food parcels each, and the needed *jerry* (or fuel) cans. Three co-drivers manned each truck; an American, an Englishman, and a German. Their job was to drive across the border from Switzerland into Germany and continue on to the assigned prison camp to distribute the food they carried.

Later, I'm told, the trucks operated out of the Baltic seaport of Rostock, thirty miles west of us, with Red Cross parcels brought in by a ship from Sweden. I suppose this is where the large number of Red Cross parcels that the German commandant had at the Flak School came from. Wherever they originated, they were welcome and were our food mainstay while Stalin stalled our departure for ten days until he finally gave up and we left for France on 8th Air Force B-17's from the 1st Air Division..

With Germany then in such desperate straits for any form of transportation, and for any source of food, I've wondered many times how the trucks and food were allowed to continue on to us without being pirated. We were told about these trucks, their cargoes, and their trips, but were given no further details.

As mentioned earlier, although we had a few hardships, we survived with the help of the Germans - not in spite of them.

Top Photo: Double Gate entrance to North 1 Compound on left where wagon is located. Compound entrance to left after gates. Buildings to right are: Nearest - guard administrative office; back - office, sergeant of the guard post.

Bottom photo: Lighter colored building to right - solitary confinement cells with covered upper window. Darker colored building building to left - guard administrative offices. In background - the tall Medieval church tower in Barth, started in 1272. The town of Barth was then a Swedish walled town. Far right - typical guard tower. Far left - beyond the left buildings- mess hall before it burned to the ground.

10
The Lecture Series

During our service with the U S Army Air Forces in WWII all young officers, flying or ground, were required when possible to attend monthly group military lectures. Even while I was a hospital patient in England recovering from combat injuries, I originally attended in my bed in the surgical ward with a group of officer patients grouped around in wheelchairs. The nurses, male and female, were excused. After I was allowed out of bed, I attended in a wheelchair. These were usually very interesting talks held or chaired by senior officers.

Even though it at first seemed somewhat unlikely, this same practice carried over to our stay in the German prison camp, but these talks, or rather, lectures were quite different. This series of lectures given by the actual military leaders of some of the key turning points of WWII gave us, the attendees, an actual window on history almost, or very soon after, it was being made. It was one of the most interesting and valuable experiences we had during our stay in Germany, although few if any non-prisoners would even have heard about it. Because we couldn't take notes, I cannot recall, after so many years, the names of the various lecturers, except those I knew very well.

We were shown the planning, execution, and outcome of many important and secret operations. Some were not known to the public until the end of the war, and some even now are not fully explained. Often with good reason or out of necessity. The lectures were held in a secure location with posted sentries. Commissioned officers, who each held a security clearance, were lectured about the details of many campaigns by the actual battle participants and leaders in front of a blackboard. The lecturers, with a damp rag to clear the board quickly if German or unauthorized personnel arrived unexpectedly, each had a burning stove at hand to destroy written notes or sketches, that gave us all the operational details of these battles. Here are some of the operations covered and their lecturer's names when remembered. The details covered are from memory only.

The Bruneval Raid

A British airborne Lt Colonel and battalion commander, whose name I've long forgotten, told of the planning and of the parachute drop on the French Channel area east of LeHavre on the night of 27-28 February 1942. The purpose of the mission was to temporarily invade a German radar installation and seize certain vital elements of the Freya and Wurzburg flak radar units, such as the magnetrons (electronic timing devices) and associated electronics. When they had succeeded in locating and safely removing the needed equipment, the group fought their way back down to the British Commando-held beach and were transported with their prizes to the English side of the Channel by elements of the Royal Navy and the Free French Navy. As our British friends said, it was like fighting your way down through the bees once you've taken their honey. Allied radar knowledge and equipment design were both vastly improved with the actual devices in hand for study of the much better equipment that had recently been perfected in Germany.

The speaker was later captured at Dieppe and spent many years in several other German prison camps before coming to Stalag Luft 1.

Even America profited as they gained the knowledge taken from these cross-Channel raids. The Wurzburg radar, the primary German radar used to aim the high-altitude flak guns, was the model, we were told, that became the forerunner of our later American SCR-582 anti-aircraft radar. It was used for many years by our Army flak units, and was later used by the U S Air Force. During the Korean War. The information gained from this ground-based flak radar unit and its associated electronic plotting tables were also used to direct close-in front line bombing by the B-29 aircraft I flew. The accuracy and adaptability were very good. It's only real limitation then was its short range of about thirty miles to "skin-paint," or get a direct radar return, or electronic reflection, of an aircraft. The Strategic Air Command used it for many years to "score" practice electronic bomb runs on simulated US targets. It was used for all SAC's bomber training up to and through the B-52, on what were called "RBS runs" (Radar-bomb-scored-runs). I made literally thousands of such practice bomb runs at simulated targets all over the US while flying B-50, B-47's and B-52's over a period of 14 years.

The Dieppe Raid

This was another cross-Channel assault, but was far broader in scope and had too many unrealistic and diverse aims. The planners, probably seduced by their earlier successes, such as Bruneval, seemingly threw caution to the winds and threw everything in "but the kitchen sink." It was also billed in American newspapers as a training exercise for the eventual invasion of Europe by Allied ground forces. If that's what it was, the results were a dismal failure, because the invasion training elements and poor results left much to be desired. Tanks were landed, then were confined to the beach by a large wall that had not been spotted by previous air reconnaissance photos. The tanks couldn't climb over it and were destroyed, along with their crews, uselessly. Most of the infantry that landed on the beaches and attacked with the tanks were either killed or captured. All of the equipment landed in France was lost. Only a relative few who came ashore were able to slink back to Blighty "with their tails beneath their legs". This mission was what the British officers called, "A bloody poor show."

A highly classified and little known sidelight of this fiasco was another successful radar station highjacking. The target was again another magnetron, or timing device, which was the "brains" of the system. It was from the longer-ranged German radar unit called a Gemma Coast Watcher. It could detect aircraft and give early warning at far greater distances than the smaller units like the intermediate Freya or the Wurzburg flak radar which was designed for close-in, precisely aimed, anti-aircraft fire by the Flak 18 or 36/37 cannons that were used to defend German cities and installations against high-flying bombers.

The British Army captain who gave that lecture had been the military escort, or bodyguard, for the top British radar *boffin*, (Scientist, to you, Yank), who came with the team to oversee and remove the desired electronics wanted. The captain told us that if capture of the scientist became imminent, his instructions had been to shoot and kill him so that he and his knowledge wouldn't fall into enemy hands. That seems strange, or was poorly informed, since the German scientists probably knew far more about radar then than their British counterparts. The former captain later re-counted this story after it was de-classified, on the TV series, *World At War.* If I see this program again, perhaps this time I'll remember to write down his name.

The Taranto Raid

The British Commando raid at the Italian naval base at Taranto, down near the *Boot*, or southern end of Italy, was recounted by then Captain Archie G Birkner, of the 82nd Airborne Division. Early in the war our brighter young officers were often exchanged with other Allied forces to spread knowledge or to gain needed combat experience. Captain Birkner was on loan to the British airborne *paras* or paratroop forces, that accompanied the Commandos. I don't now have much more information, except that the Allied forces caused great destruction at the large Italian naval base and to its shipping. It was also another bloody mess with very high casualties on both sides. Very few of the attacking force, if any, made it back alive. Most were killed or captured.

Captain Birkner, who spoke fluent German, was a *spit and polish* officer in his immaculate jump uniform and highly polished jump boots. He was greatly respected by both the Allied and German officers at our stalag. This was further enhanced, no doubt, by Germany's well-deserved respect for its own Luftwaffe parachute troops. They were among the best fighting men in the world, as was shown at the battles they fought in Belgium at the beginning of WWII, with their leader Col Bernard Ramke in North Africa, and later at such places as Monte Cassino in Italy. For quite a long stretch they gave Allied attacking forces a "bloody nose" while stopping our forces for months.

Captain Birkner's assigned camp duty, was officer in charge of Red Cross ration distribution for the entire camp. In this duty he was permitted to wear a purple parole armband with a white cross and walk, in uniform, unescorted, back and forth to the town of Barth to handle needed details about food deliveries that arrived at the small railroad station.

The Battle Of Arnhem

You may not remember the battle for the bridge at Arnhem across the Rhine that led to Germany's gateway, but it was the last bridge in the three-bridge battle plan for what was called later, "The Bridge Too Far". This plan was an overly optimistic mission that was covered, though not very well, in the movie of the same name. If successful it would have opened the gates for the Allied ground forces to flood out into the plains of northern Germany and begin the race to Berlin - and the end of the European part of WWII in 1944. When the effort foundered the war dragged out the better part of another year and allowed Russia to grab and hold much of Germany for forty years.

It was another of the many failed wartime operations, similar to Field Marshall Montgomery's failure to close the gate and prevent the escape of seventeen trapped German army and *Waffen-SS* divisions at Falaise in France, General Bradley's "bloody nose" in the original attack on the almost vacant Siegfried Line, Patton's disastrous raid at Hammelburg, the Dieppe landings, and many other failed operations, large and small, that foundered, not because of the ability and determination of the troops involved, but because of very poor intelligence work, planning, and the lack of proper reconnaissance, that made the later *Battle Of The Bulge* possible.

This shows up in many of the post war histories. The cross-Channel invasion by Allied troops in June 1944 is only another very bad example of poor planning, despite the heroic efforts of our troops at the scene. Among many other failures, Ranger assault troops were sent to scale vertical cliffs to destroy large naval coastal-defense cannons above the beaches, although these big guns had already been removed long before for use in another location. Aerial photos would have shown the current situation, but were not taken. Many tens of thousands were killed or wounded, because of out-dated aerial photos and the ineptitude of the planning and/or leadership personnel responsible.

At Arnhem, unknown at the time, were two German SS *panzer* (tank) divisions resting, refitting and re-arming nearby after returning from bruising Russian campaigns. They were in that same area, but were again unseen by the complete lack of aerial photos. They were the 9th SS Panzer Division, *Hohenstaufen* and the 10th SS Panzer Division, *Frundsberg*. These two divisions, although under strength, were breaking-in new armored vehicles, and were engaged in training very young, green replacement troops, but they were still strong enough to turn the tables against the lightly armed British airborne troops. The German units bull-dog like tenacity put a "thumb in the dam". Proper aerial photo-reconnaissance would have prevented this loss also.

The British parachute battalion commander who led his unit from the drop zone fifteen kilometers west of Arnhem into the town located on the north side of the bridge was our speaker, He created a chalk drawing on the blackboard showing the landing ground, the route into the town, the various obstacles along the way, and the bridge location. Having landed so far from the objective, but where he was ordered, he lost many of his men and most of his equipment before getting to the town or bridge. A lightly armed parachute unit with only the weapons they can carry has very little chance against armored vehicles and artillery, no matter how well they fight. Very few of these men survived.

In the fierce fighting in the narrow streets of Arnhem, the lieutenant colonel received a head wound that left him unconscious and bleeding in the street. Townspeople dragged him into a cellar so a local doctor could treat him. This saved his life, but soon after he was found, given first-aid by the Germans medics, and sent to a hospital in Germany. After he recovered, he was moved to our stalag's West Compound.

This failed operation which might have shortened the war by many months, instead led to many more tens of thousands of Allied casualties and vastly more among not only the German troops, but among the civilians in the countries of Europe.

The Doolittle Raid Against Tokyo

Lt Colonel, then, Charles Ross Greening, CO of Group 1, or North One Compound, was its lecturer. As one of the pilots on this raid, Ross detailed the entire mission in great detail from the selection of pilots and crew members, to the training program needed to teach the pilots how to fly the light twin-engine B-25 Mitchell bomber aircraft from the small deck of an aircraft carrier. For that purpose an aircraft runway was marked by paint stripes to simulate the carrier deck space limitations. The crews practiced the short-field take-offs until each was capable of getting into the air safely. There were no aircraft carrier catapults back then. Even if there had been, the B-25 was not designed with a strong enough landing gear or fuselage needed to use one. Ross described the loading of the planes aboard the aircraft carrier USS *Hornet* by crane, the life aboard on the long sea voyage, and then the hurried takeoff when the carrier was spotted by a Japanese fishing trawler which might have sent a radio warning back to Japan.

It wasn't known if the vessel was actually a picket ship standing guard to warn Japan of an enemy's approach, or if it's purpose was simply fishing. The range of the radio equipment probably aboard, was also unknown, and finally, there was no way to tell if the enemy vessel had even sent a radioed warning back to Japan before it was hurriedly sunk by naval gunfire.

In light of these possible dangers to the carrier group, which could have been attacked by land-based bombers if spotted this close to the Japanese home islands, the B-25's were ordered to take-off while still 800 or more miles from Tokyo, instead of the planned distance of 400 miles out. This added a sudden new fuel-versus-range problem that would prevent the B-25's from reaching planned airfields in China.

Ross detailed the mission flown by his crew, from the long over water flight to Japan, the low-level bombing attack after they spotted their assigned target, the crossing of the Japan Sea, and then the run down the China coast to their turning point inland to the mainland of China. As they headed west in the gathering darkness the fuel warning lights finally came "on" to wave good bye. Each man in turn jumped out into the darkness. They landed in the Japanese-held territory of China. The Chinese people at great risk helped the crew to find each other and to walk with their guides toward safety.

In all it took Ross and his crew almost six months to walk across China and finally return home. During the trek they were constantly hunted by the Japanese and would have been killed if captured.

The raid did little real damage, he told us, but the Japanese people probably began now to see a fearful hint of their future. Their imagined safety had been rudely shattered. The raid's even limited success also gave a great boost to American morale, which up to this point had witnessed only many defeats, disappointments, and losses. Only the broad outline, but none of the details of the Doolittle were released to the American public until after the end of WWII. Here, we had the whole story.

Ross was the inventor of the simple and inexpensive low-altitude bombsight that each of the planes used. He also described the "fake".50 caliber twin machine gun mount added on the rear of the planes. In that model B-25 a real turret and gunner, or even real guns and mount, would have been much too heavy for the plane and gun location The wooden "guns" and mount he created weighed very little and did not affect the planes' weight and balance. The black paint used made them look real enough, hopefully, to discourage enemy fighter pursuit. As it turned out, no fighters chased his plane.

Although this wasn't part of his lecture, the following paragraphs explains how Col Greening became our Group One leader, or North Compound One commander. It's from his book, *Not As Briefed.*

Later, after Ross's return to the US from China and the Doolittle raid, he was sent to Italy as the commander of a Martin Marauder B-26 bomber group. Ross was forced to parachute from his flak damaged aircraft while bombing a target in the Naples area. Afterwards he was captured and taken to an Italian-run prison camp. He later escaped from that camp, with others, and hid out with Partisan's help in several mountain cave hideouts. During almost six-months, while he hid with his companions and received help from Italian families nearby, he used many hours doing "combat artist" sketches.

The enemy finally found his group, killed the Italian resisters, and shipped Ross, after he was identified as a American airman, through the POW system to our Stalag in Germany.

Lt Colonel Charles Ross Greening,
CO of Group One - North Compound One.
Flew with General Doolittle on Tokyo raid
from aircraft carrier USS *Hornet.*

Typical Barracks in Stalag Luft 1
Illustration by Joseph Reus

11
Prison Camp Life

After capture, the travel by car and tram on route to the Luftwaffe interrogation center at Oberusal, then the seemingly endless four-day train haul straggling across Germany in a cold and dirty boxcar as the lowest priority freight, we had finally reached our destination.

It was not an old multi-level Gothic stone pile with barred windows as some had probably envisioned, but a normal air force camp with pre-fabricated one-storey barracks arranged in a roughly square setting grouped around a parade ground. Although we were enclosed within ten foot high barbed wire fences with tall searchlight and machine-gun equipped towers, we were now in a safe backwater. Life was new and somewhat strange, but not foreboding. Having grown-up with German speaking grandparents, who spoke American-English with considerable accent, I probably felt less strange and more "at-home" than most of the other new prisoners with different ethnic backgrounds.

To the east, across the open farm fields, then the small bay, we saw the red tile roofs of the small fishing village of Barth with the prominent spire of its medieval Swedish church. I imagine that all who lived at our Stalag still have that church spire and village in their mind's eye. Not far from the old church there is a huge stone gateway that in medieval times had been the western gate in what was back then a Swedish walled-city. Frederick the Great of Germany captured this area when he conquered all of the southern shore of the Baltic Sea and incorporated it into what is now modern Germany.

The word *kriegie* was a shortened version of *Kriegsgefangener*, or war prisoner, for which this book was named. You'll see it in many uses, because Americans servicemen create slang names or words for convenience. The Germans did the same. Many words invented by them have become well used all around the world. As an example, *flak* (fl-a-k)is an acronym made from three words, *flieger-abwehr-kanone* or *flyer defense cannon*. Another is *Gestapo*. Again, it's three words. *Ge-sta-po*, or *Geheim-Staat-Politzei* - which means *secret-state-police*. There are many such usages, just as we have.

When we arrived, the areas to the north of the camp were open fields and forests that were later to be covered by, first, the barracks of North Two compound - and finally, those of North Three compound. As time went by the camp enlarged to shelter ever greater numbers of Allied prisoners from the steadily increasing air war and finally, near war's end, by the addition of the prisoner evacuees from eastern camps that had been emptied shortly before being overrun by the rapidly oncoming Russian hordes. The tall evergreen trees beyond those fields also were later lost to our view being blocked by the many hastily erected pre-fabricated barracks. The view to the west also was of tall pine trees, but within that forest area there were various trails that led to the western branch of the body of water called the Boddensee. It surrounded the peninsula on which our camp lay and encircled it to connect with the bay on the east at Barth and led out eventually into the open Baltic Sea.

In the following Summer, that of 1944, we were permitted to sign up and leave our compound on parole in groups of 200 or so, to walk westward through the forest to the Boddensee and swim under the watchful eyes of our regular escort guard in his small rowboat. I spent almost the whole of each period swimming from one limit to the other, back and forth, to help strengthen my arms both of which had been broken when we crashed into the English Channel during the previous Winter. Since our escort was unarmed he was more like our protector and guide than keeper. Because he spoke good English and was friendly, he was also a good information source. He and I got on quite well. He told me of his former years as a merchant seaman in the 1930's and of his trips from Hamburg to the port at Baltimore near my hometown. He told me that he'd always been treated well there and hoped to go back again. We also later in the Fall had a few more hikes under parole that took us into the forest at the north end of the peninsula. Later as conditions worsened and the camp became much more crowded we lost these pleasurable excursions. Also I never again saw my German friend. Being able-bodied he was probably sent to the East Front to fight as an infantryman with a German air force Luft-Land division (air-landing or airborne) in the last-ditch defense.

Soon after I arrived, I remember listening to an RAF pilot who had been a prisoner for several years, who said he could tell what airplanes were flying over us at night. Whenever we heard aircraft engines, he'd say, "Those are *Lancs* (Lancasters). Maybe later he'd say,"Oh, Oh, here come 109's." I was skeptical and didn't really believe him at first, but as time and Pow months went by, I, too, could began to identify various aircraft engines by myself and knew that he had been correct.

Soon, also, during daylight raids we could easily tell the difference between B-17s and the B-24s very different engine sounds before we could make them out visually. No British fighters had the fuel range to get into Germany as far as our camp on the Baltic, so we didn't learn to identify them. However, the twin Allison engine powered American P-38 Lightnings and the later single engine P-51 Mustangs were also easy to recognize by the sound of their engines. I don't really remember ever seeing or hearing P-47 Thunderbolts.

German aircraft sounds, especially the multi-engine types were much different from ours, because their engines were not synchronized and gave off a beat-beat-beat. Our engines were synchronized and were quieter and more easily identified. In later years during trips to our old airbase at Tibenham I was able to hear the sound of the Rolls-Royce *Merlin* engine's similarity between the Spitfire and the P-51 Mustang.

Later during our stay vast air armadas of the 8th Air Force flew through our area on local raids or on the *shuttle* raids that bombed German targets then landed at Russian-occupied Polish air bases. On subsequent days the refueled and re-armed bombers returned to England after bombing German targets along the way back. The great continuing roar of the huge passing bomber fleets was a sound we'll keep in memory until we die, but never hear again..

The British Mosquito was in a class by itself. With its two big Rolls-Royce *Merlin* engines at full throttle it made a wild scream like an enraged panther, as we would learn a bit later during several ferocious, extremely low-level, high-speed ground attacks in our immediate area over the airfield and an attack against an FW-190 above the bay at Barth.

Barracks

After arriving at the camp our long line of new prisoners was marched to a hall where we were given a briefing by a young Luftwaffe captain. Next we were separated into one smaller group that marched away to West compound and the much larger group I was with that was assigned to the then new North compound. Another much older Luftwaffe captain, who'd flown during WWI, gave us his briefing before marching us to and through the double gates of our newly assigned North compound. We were next taken to the shower room and de-loused' with kerosine (Germany didn't have our DDT, ouch - sorry). After that we showered as quickly as possible to remove the skin-burning fluid and again dressed in the same dirty clothes which could almost stand alone.

After we re-formed we marched out to the parade ground. In turn our group of twelve was "counted-off" to an assigned room, #7, in barracks, #7, in almost the middle of North Compound. Not being bashful, I marched in and claimed the lower bunk on the northwest corner of the room. A co-pilot from my 445th Bomber Group, Lt Kenneth Covey, who flew with a different crew and squadron, took the upper bunk in the 2-bunk stack. He and I remained in this order throughout the balance of WWII, even though after several months we moved to room #7 in barracks #10 in the same order.

Our new home had previously housed German Air Force cadets, undergoing combat training. The door to each room originally had on its lower front panel a painted coat of arms of a revered and ancient Teutonic warrior to inspire these young men. They were later painted out by us.

The normal pre-fabricated, single-storey, German Wehrmacht (the word means Armed Forces - not Army) barracks had six large rooms along a central corridor with two smaller anterooms at each end, plus two latrines. The main rooms of 12 feet wide by 15 feet long normally held twelve men, but during the Summer and early Fall of 1944 each room had to take in an extra three men until a new compound, North Compound Two, was completed by the German carpenters and their Russian air force prisoner helpers and was ready for occupancy. The smaller end rooms usually held four English NCO's. The steady and very large influx of new kriegies during the Summer of 1944 even required tents to be erected temporarily around our parade ground and in-between our barracks to hold the growing influx until permanent space could be provided in the new buildings. Generally, a barracks held about 88 men, and a compound around 1000 men, although it varied widely from time to time as needed.

Byerly's Revolt

During the early days after our arrival at Stalag Luft 1, during the period when Col Jean R Byerly was the Senior Allied Officer, or SAO, we had a test of wills between Col Byerly, a former bomber group commander and the German stalag commander, German Army Oberst (Col) Scherer, which we called Byerly's Revolt. It came about when a 1st Lt refused to obey an order from a German security police NCO, relying on the words of the Geneva Convention agreement of 1929, which said that officers could be given orders only by someone of similar or higher rank. The Lt was promptly arrested and taken under guard to the German camp adjutant. Since the Lt wouldn't back down he was given an eight days sentence in the klink, or solitary confinement. Col Byerly put on the purple armband, signifying that he was on parole status, and stormed through the quickly opened gates and marched down the road to the camp headquarters. There he told the kommandant that our men would not fall out for evening roll call until the Lt was released, before stomping back to our compound.

Oberst Scherer, being a man of few words, sent a company of Luftwaffe personnel to line up along our compound's east fence. They wore infantry gear and were equipped with rifles and machine pistols. The Oberst sent a message that said in so many words that roll call would be held at five o'clock, and if our men were not in place, he would order his men to fire across the compound at five minutes after.

Col Byerly, having gotten much wiser very rapidly, ordered "All right, men, "Fall Out." The tempest in a teapot ended as fast as it had begun. To show good faith and sportsmanship, Oberst Scherer ordered the Lt released from confinement the next morning.

The Allied Camp Staff

The original Senior Allied Officer (SAO) when the first Yanks arrived was a British Royal Air Force Group Captain (Col) Middleton. When the Yanks outnumbered the RAF by an ever-growing margin, USAAF (U S Army Air Force) Colonel Jean R Byerly became the new Senior Allied Officer. Later when USAAF fighter-pilot *ace* and group commander, Colonel Hubert L Zemke arrived, he became SAO, since he out-ranked Byerly. "Hub," as he was known, wrote a very accurate book about his experiences as commander of our camp. It's called, *Zemke's Stalag*. I have a copy. It's well worth reading.

Lt Col Charles Ross Greening, the Tokyo raider, was named Group One Commander and Commander of North compound on arrival in January 1944. He had commanded a Martin B-26 Marauder medium bomber group which flew in Africa and later Italy until he was shot down and captured near Naples. When North Two Compound was completed in the late summer of 1944, we were re-named North One compound. As time went by and the camp rapidly expanded each compound was assigned an American commander appointed by the SAO with the needed staff officers to cover normal administrative functions. The German camp command did not interfere with our administration.

Appell

Our first camp activity was run by the compound's social director, or German compound commander, a *Luftwaffe* major. He, with his adjutant and NCO staff, had us fall out and join formation on the large central parade ground by barracks or squadron formation for *appell*, or roll-call. The British term is *parade*. The team of German NCOs fanned out and counted the number of men in each barrack's formation, then came together to add the totals. Added-in were the sick in barracks and known absences, such as the *klink* (solitary confinement), and the sick at the clinic. At that early date we did not yet have our own klink, but had to share the only one, in West Compound before we became too unruly and needed more cell space.

Older German door guards were positioned at both entrances of each barracks and at each gate during appell until the formation was dismissed. If the numbers agreed, the German major exchanged a salute with our Senior Allied Officer and Group One CO, then marched off with his men.

The Senior Officer present or his adjutant made any necessary administrative announcements and then dismissed the formation in a formal manner. We soon learned that the more formal we were in our military functions the greater respect we earned from our captors. This protocol is very important in handling large military groups during adverse conditions such as prison camps. Our use of military customs was strictly followed throughout our stay at what we later called, "Barth on the Baltic". We even were allowed to erect a sign over our entrance gate with that name.

This appell ritual became part of our daily life, as we were counted before breakfast and before dinner throughout our stay, from arrival until the German command turned the camp over to American military control before the Russian forces arrived.

Though I didn't think ahead at the time or keep a tally, I now conclude with a navigator's accuracy that I stood roll-call 781 times in all of the weathers that the Baltic Sea coast can provide in each season. Of course, I've taken off the 16 days while I was on vacation in the klink. We never had tornadoes or hurricanes that far north, but we had rain, snow, sleet, and every form of frozen pellet and Summer rain and fog that nature could muster. When the weatherman "screwed up" we even got some sunshine in the beautiful Spring and Summer of northern Germany. The German staff also had to put up with the same vagaries of nature to the exact degree that we did, and in a sense, they were also prisoners to their duty, but they were able to go home or to their barracks at night. When you think about it, a jailor's routines do not stray very far from that of their captives. I've read that there were more than 40,000 American prisoners in Germany when the war finally ended.

On one snowy, cold and overcast Sunday morning while the German NCOs were totaling up their numbers, and the POWs were waiting in place to be dismissed, Lt Alfred Sheppard shook the hand of his roommate, Lt Eugene "Sabertooth" Johnson. Instead of releasing each other's hand, they appeared to be stuck together. The more they pulled and twisted, the more they got their arms and finally their legs entwined. Finally they fell to the snow-covered ground, tangled in arms and legs. A very old guard at the nearby barracks door behind them propped his rifle against the barracks wall and came forward in an attempt to help. As he tried to untwist the two men lying on the ground, his hands were grasped and he too became part of the pile and also was soon kneeling in the snow. At this point the German compound commander saw what was happening and got the picture. He walked over, said nothing, but glared down at the pile of men. The old guard saw his boots, looked up and went "beet red". He'd been "had". It was an old vaudeville gag. The penitent guard scuttled back to get his rifle. We'd all silently watched. When it was all over we all applauded, loud and long. Imagine the sound of the whole compound clapping in unison. I guess that most of the Germans present also had great difficulty in stifling their laughter and keeping a straight face.

Several notable names also lived in barracks #7 then #10. One was Lt Bernard Barker, a Cuban-American B-17 bombardier from the Miami, Florida, area, who came to newspaper notoriety, or ill fame, much later as a member of the "Watergate Gang" with G Gordon Liddy's group, which enabled the Democrats to stage the *coup* that removed Nixon from the presidency. Another man was Bob Hoover, a former Spitfire fighter pilot. Yes, Americans, though in the USAAF, flew Spits in the Mediterranean area, because we had no combat-capable fighters there. Later, for many years he was one of the top American and world aerobatic aces. He was also the *chase* pilot for Brigadier General "Chuck" Yeager during the flight that first broke the sound barrier of Mach 1 out at Edwards Air Force Base in California. Bob was also very well known in Europe for the air shows he put on during WWII in many types of aircraft, both in Allied and captured German aircraft. He was and still is a legend among aviators.

Not long after our arrival on 1 March, Easter arrived on a cool and cloudy early April Sunday morning. By pre-arrangement with the German compound commander, after the count was completed at morning roll-call, and before the formation was dismissed, our trumpet player, Lt Mike Spodar, was permitted to play his beautiful rendition of the *Easter Parade*. Unlike most trumpet players, he could carry the whole piece, not just the solo, and do it beautifully. Mike's sentimental music really set the scene. I don't believe either of the two German officers present, or the group of *Luftwaffe* NCO's and guards with them, knew the significance of this song to young Americans aviators. Even if the song was known in Germany, the American solemnity was so obvious, and because of the many damp eyes, their faces were as solemn as ours. It was a moving moment. I actually believe that this tiny event created so soon after our arrival established a small bond of shared humanity between us, even if we didn't realize it at the time and were wartime opponents.

Laundry

Fire buckets made excellent washing machines and were in constant use. You simply filled them with warm water heated on the room's tile stove. For agitation or movement you used an empty Klim can with holes punched inward (so you wouldn't tear the cloth) with that handy stolen nail you've hoarded for all your small drilling needs. After attaching a wooden handle about two feet long, probably made from a part of a bed slat, and after adding shaved soap to the water, you were ready to wash. While you read a book with your left hand, you pumped up and down gently with the right hand, The handle pushed the Klim can up and down in the bucket and squished soapy water out of the holes in the can until your arm got tired and your clothes got cleaned. An interesting book saved you from boredom. It was also much quieter than more modern washing machines - and didn't take any longer to use. Sorry, no dry cleaning or ironing.

Some of the more enterprising men went into the laundry business, using two buckets - one for each hand. I wonder if they ever received the promised payment of the *I.O.U's* they received for their services after liberation. A rewarding exercise session in any case.

Hobbies

Almost everyone had hobbies of one sort or another, from the useful to the outlandish. Most of the young flyers at the camp had probably shown their earliest interest in aviation by building models of aircraft when they were boys. Now some of the men built flying models while others constructed scale models for display. They acted no differently in prison, even though the raw materials were not available in hobby stores and were much harder to find. Being bright and inventive they built first what they needed, then what they admired. In our room, the first item made and completed in only a few hours, was a deck of cards, so the few fanatics could play bridge and later cribbage as a mild form of combat. Yes, that's what all such, so called games, are about.

A neat grid was burned with a hot wire into the surface of one of the two wooden tables so we could play checkers using round, water- colored, cardboard cut-outs. They started out with one side being darkened by pencil lead - and the other cardboard colored. The chess sets didn't come until somewhat later after the makers were able to acquire pocket knives for whittling. Some pieces were so crudely made that there were heated arguments about what a certain piece represented. To end the battles, those pieces had to have a printed letter identification under their base so the "combatants"could agree on identification. It, too, added an extra sound level to the room.

Trade

We were permitted to have pocket knives for carving or eating with blades no longer than 3 inches. They could be bought from the Germans, but you had to buy them only through your compound's American, German-speaking, official *trader*. This was strictly enforced and prevented inflation of the price. Our *currency* was the American cigarettes, whose names you probably remember, such as *Camels, Lucky Strike, Chesterfields,* etc., that were provided in five packs of twenty each in the eleven pound Red Cross food parcel. Each of the cigarettes was worth one German Mark. My pocket knife cost only three American cigarettes, or three Marks. Many other needed items could also be gotten this way. American cigarettes were greatly in demand, as we quickly learned. Many, if not most, of the items and supplies used in the art school and other training programs were bought, legally, with cigarette currency. Trading was probably used to support many other classes and much more than I observed. Books usually were gifts from student unions or groups in neutral countries primarily in Europe, but also from South America. They were in many languages. We also received large boxes of books from England and American, hard-bound and paper-back sent through neutral counties.

I believe the radio parts that were assembled into receivers to bring us the BBC news from England were bought with Red Cross supplied cigarettes. I have no doubt that the first radio, in West compound and used by the RAF long before we arrived, was also constructed from parts made available in this way. There likely were more things bought than I could possibly know.

One of the friendly German guards, who could speak passable English, even had a joke about the German cigarettes issued to the guards. He said that before the war German cigarettes were made with half horse manure and half tobacco. Then he said when the war came they left out the tobacco. Their humor was very much like ours. We laughed at the same things and just as often.

Magda Goebbels, wife of the propaganda minister, had the tobacco and condom monopolies, we were told, and became immensely wealthy, though it didn't do her much good in the long run. Her last "apartment" and final resting place was in Hitler's bunker and that didn't even have a window. Her window into another world came only from a poison pill.

The German guards grew their own small tobacco plants with tiny leaves in window boxes at their barracks. When the leaves were picked they were threaded onto a string and dried in the Sun. We could see the hanging tobacco leaves at their barracks windows, next to the small rabbit hutches they kept to provide more meat. We saw them when we walked past down to the medical clinic in the West Compound. After slicing the dried tobacco leaves with a razor blade they could be smoked in a pipe or rolled into cigarettes. Cigarette butts were never thrown away by the guards, They put them in their pockets and smoked whatever was left in their small pipes. I don't know if they picked up cigarette butts discarded by the Americans, but because tobacco was in so much demand, and because American tobacco was so desired, I wouldn't have doubted it. Being a non-smoker I was "rich", but there was very little to buy.

We had a small PX or Post Exchange in an anteroom at the front of our mess hall. It was used primarily to trade food items from the Red Cross parcel you didn't want or like, but could be used to trade for anything else on hand. You paid the dealer two cigarettes as a transaction cost. Items already on hand also had a quoted price in cigarettes which went to the seller. You placed a cigarette price on what you wanted to sell. The biggest problem was that if you were selling a food-can you didn't like, it was probably disliked by many others, and did not sell readily - such as Liver Pate, which few wanted.

German Tobacco

There was a German tobacco "ration" for us, too, when supplies were available from our captors. It consisted of two cigarettes and one tiny "cigar" per day. The NCO ration was two small cigarettes per day. I don't know if these were bought with our funds or were provided by Germany, but I believe they were supplied by Germany. They were so far below our standards I can't believe our command would have bought them. In any case, these smokes were so poor only a masochist would actually smoke them more then once, and the first time was only experimentally. However, since we tried to put everything to good use and waste nothing, never turning down anything that was offered, we used the cigarettes for fumigation to kill the gnats and mosquitos from the nearby marshes along the Boddensee banks that gradually accumulated in the room during the warm months.

In the balmy late Spring and Summer when these flies and gnats gathered we used the ubiquitous powdered milk Klim cans as burners. Many holes were punctured in the can and scrap paper was covered with shredded German *Elegantes* cigarettes heavily spread over it. The paper was lit, the top was put on the can, and the tobacco was left to smoulder. We'd all leave after putting four or five of these cans around the closed room. This filled the room with enough smoke to kill the insects in an hour or so. Then on return we had to air out the room by fanning the door and the large wooden outside shutters to clear the smoke. Fortunately we had been deloused from any "critters" we might have had when we arrived, and we were able to keep clean enough so that no lice infestations were brought in later. I never saw a mouse or rat. I suppose they were afraid to come near us. In Asia they're considered food.

Another tale had to do with a cigarette "fiasco". While Erwin Rommel and his Panzerarmeeafrika were fighting across the deserts of North Africa trying to capture Egypt and the Middle East oil fields, the German propaganda ministry had all of the tobacco collected so it could be re-packed as *Cairo* cigarettes for a big propaganda campaign. When Rommel was stopped, then turned back, at El Alamein, these cigarettes couldn't be released. Then the Allied propaganda would have trumpeted their failure. So, it was back to the factory for re-packaging the cigarettes with a different name before release. In the meantime, due to the time lag involved, this caused a severe shortage of tobacco for the troops and German civilians. This was probably another factor in the demand for American smokes. From what I remember from that postwar era, the cigarette lag lasted for years in Europe.

The Art Committee

After we'd settled down a bit I made a simple set of drafting instruments. The dividers began with the end points from two safety pins. The ruler started with an arbitrary inch beginning and took the shape of a Weems navigation plotter. It had a center protractor for angle measurement and an edge scale broken down to one sixteenth of an inch. It was made with great care from a thin piece of wood "liberated" from the barracks. It allowed me to do precise drawings without a T-Square and tacks. I couldn't tie up the limited table space by tacking down a drawing. By using a baseline at the bottom of a page - and erecting vertical lines at either side - I drew more accurately than with a wobbly homemade T-square. I did not make a *french curve*, but I did make 30-60 and 45 degree triangles.

The word soon spread around that I would help with drawing plans for model planes. With the aid of various pilots and aircrewmen who had remembered dimensions and shapes, I could select a scale that would fit the page and help establish correct proportions. Together we were able to piece together what was needed for the plane that was wanted. It was a continuously improved process that lasted throughout our stay at Barth. There were P-51, P-47, P-38, Spitfire, Hurricane, B-17, and B-24, B-25, B-26 plus many other types for which drawings were made, even British. I was surprised at how much data we gathered. My only payment demand was that I receive the needed paper. It was always in short supply. A few drawings were in ink, but most were from the pencils that had originally been supplied with our Red Cross suitcases at Oberusal. Pencils were always available from the cigarette trade.

Not long after we began model design drawings, Lt Col Ross Greening, our North Compound commander, stopped by the room. He told me that he'd heard what I was doing and looked over my "tools". At first he asked me if I could teach others to make similar drafting instruments. I assured him that I could teach anyone who wanted to learn. He added that he was about to start an art class and invited me to help as the drafting instructor. I was glad to join in. As Ross' *assistant*, I taught the making of the needed drafting implements and then how to use them. Now I'd be able to get more paper, pencils, pens and ink, and establish a new hobby that would fill many satisfying hours.

The committee's real purpose other than art classes, unknown to most of the students, was to create the map and document section of the escape committee. The Germans allowed us to buy the needed paper, pens and ink, brushes and paints, and many other items we needed for all of the types of art classes that grew from it. They probably thought that if we stayed busy we'd stay off the fences and out of trouble. It probably was something like - out of the frying pan into the fire - for the Germans.

Gradually we got into the map tracing and making of other items I've mentioned in the *Escape Attempts* chapter.

To keep this material out of sight, some of what I worked on was done in the room that Ross shared with Lt Col McCollom, the G-2, or intelligence officer, on the staff of the Senior Allied Officer, in Barracks #9, the *head shed*. As we worked together, I found that Ross, with his four-year art degree, could work and teach in almost any art media. If you have seen the two books, each called *Not As Briefed*, that contain the combat artist sketches done by Ross you'll agree. The second edition is far more comprehensive than the one he published at the end of WWII.

As time went by I also got to know a little more about Ross' background. He'd gotten his commission under the Thomas Act's provision that allowed people to request a commission if they had a four-year degree, even if it was an art degree. He applied, was accepted, and went to flying school at Randolph Field in the mid-thirties.

I have a copy of Ross' original combat sketch book from 1945 plus his later more complete edition of, *Not As Briefed*, filled with WWII planes and characters.

We met several times after the war. The last meeting was at the annual *Doolittle Raiders Dinner* in 1955 at the officer's club at MacDill AFB, Tampa. I was there with my B-47 crew to attend the annual Evaluation School for Strategic Air Command Lead Crews. Until his untimely death in 1955 at Christchurch, New Zealand, I'm told he was the Logistics Officer for American Antarctic missions.

Field Day, 4 July 1944

For the celebration of our 4th of July we planned to have an entire day of track & field, boxing, model and art exhibits, softball and volleyball games, plus a yacht regatta - open to not only the Kriegies, but also to the guard officers and NCOs as guests and spectators. Most of the events need not be described, being well-known enough, even though they were humorously approached, especially the slap-stick wrestling parody put on by Sheppard and "Sabertooth" Johnson - grunts and groans mixed with high-pitched giggles. You didn't have to understand English to enjoy their buffoonery.

The yacht regatta, of course, required advance planning to set vessel specification and limits as to size, and materials, allow time for vessel construction, to check meteorology, and to decide on the size and location of the waterway to be used. Since the bay between our camp and the town of Barth was probably too large, we settled upon a lengthy puddle at the east end of our parade ground.

The vessels were restricted to 6 inches in length with single sail masts not to exceed 8 inches. Keels were to be made from used razor blades. The naval committee was under the supervision of a captain, since he sounded a bit like Navy. The races were held with a mild west wind. Unlike most outdoor sporting contests we prayed for rain, so the puddle would be big and deep enough for the regatta. Lacking that we'd have to fall back on the bucket brigade.

Those seen blowing the sails of their yachts would be summarily disqualified. A jolly time was had by all, even the landlubbers.

A Drawing Assignment

One rather novel happening in the Fall of 1944 was a request from one of the Luftwaffe captains that I do a set of twelve color drawings of German aircraft for him. I have no idea where he heard of, or saw, what I was doing. The first package I'd received from my father in Baltimore contained, among other things, a requested box of 24 colored pencils. I told the officer that I'd be glad to do the drawings, but I'd have to get permission from my compound commander to do so. As soon as I could, I asked Ross if I should go ahead. He said, "Sure, but get paid in food, not money."

The captain came back with a photo album that was laced together with a red cord. It come apart easily so that I could remove the pages which were made from a heavy, cream colored, water-color paper. To give me information about the twelve planes he wanted, he also brought several copies of the German Air Force magazine, *Der Adler* (*The Eagle*), and a group of cut-outs from magazines.

Max Schmelling

I gradually completed the drawings. One late afternoon while I was doing a drawing of a Focke-Wulf FW-190 fighter plane at the table in our room, the door opened and in walked a group of five or six officers. Ross was in front, followed by Max Schmelling, the former world heavyweight boxing champion. Behind him was the Senior Allied Officer Colonel Hubert (Hub) L Zemke - former CO of the famous P-47Thunderbolt, 56th Fighter Group, several of his staff officers, and the barracks 10 commander. Strangely enough, none of the Luftwaffe officers or guards was in the group that entered the room. Our hallway and open door quickly collected a crowd when the word was passed around about our new guest

Schmelling saw what I was drawing, walked across to the table, picked up a red pencil and autographed the upper corner of my drawing. What I most remember is his huge hand, almost as big as a bear paw, as he reached over to shake hands with me. He spoke English fairly well as we talked. He told us that we were all sportsmen who had gotten along well in the past regardless of what country we were from. He said the war would soon be over and we'd quickly go back to being friends again; all in the name of sportsmanship. The visit was over in a short time, and ended as he marched around with our officers to see a few other parts of the compound. Then he headed back to the gate, shook hands with his escort, and walked to his parked limousine. Afterwards, the word quickly made the rounds that the war actually would be over soon, that the Germans were losing, and this was *Plan B, We are your friends*. Some said he'd come in a big, black, Foreign Office car with fender flags, but I didn't actually see it. We had heard earlier that Hitler, with his white racial supremacy theories, had gotten very angry with Schmelling when as a white man he'd lost his boxing championship to a black man, Joe Louis.

Max had trained as a *Fallschirmjaeger*, or paratrooper, with the Luftwaffe. He supposedly was injured during the fighting at the battle of Crete in the Mediterranean Sea. One guard NCO, however, told us later that Schmelling had actually been injured in a drunken driving accident in a German Kubelwagen, or jeep, and had to be released from active duty due to his injuries.

The German captain later collected the aircraft sketches after I completed them, plus his photo album. I don't remember if he commented about the unrequested autograph on the FW-190 page. He brought us food that he had bought in the small town of Barth, which we shared both then and later.

Craftsmanship

There was every sort of craftsmanship practiced. It seemed that you could always ask around and find someone to show you how to do or make what you needed. It might be a Yank, a Brit, or even a helpful guard. Among many other things, the guards showed us how to re-sharpen our double-edged razor blades in a glass of water as they had to do.

People with experience led or taught, regardless of rank Others volunteered to learn and fill the needs. We had tailoring that was later used in escape attempts and probably many more crafts than I'm aware of. As an example, men whose family were shoemakers in America, and who had some knowledge or experience, volunteered at the shop which repaired our shoes. Tinsmiths formed a group to roll-out the empty tin cans to make baking dishes and other tableware not provided or available. Scissors were used to cut-out the insignia of rank required for officer's wear by the Geneva Convention. Insignia was cut from cans and had nail punched eyelets so it could be sewn to the uniform. The yellow colored inside of a can was used for second lieutenants and majors - the silver colored outside was used for first lieutenants, captains, lieutenant colonels, and colonels. The NCO's had their chevrons stenciled onto their jackets sleeves neatly with black paint.

Since I'd been issued an RAF uniform when I went through the transit camp at Oberusal, I embroidered non-rusting second lieutenant bars on the jacket. White thread from the Red Cross sewing kit dyed in tea provided the almost gold color needed. For nationality I embroidered a small American flag on the left shoulder, plus navigator's wings above the left pocket. Because of my strange "get-up" in an American barracks, many of the *Yank* fighter pilots asked if I'd been with the American Eagle squadron that flew with the RAF early in the war. They must have been near-sighted. This was even more pronounced when I cut off the four-pocket tunic to create a British-style battle jacket with sewn-on belt.

Camp Duties

The pilot, Pete Abell, with whom I'd jumped into Germany, was given the duty of distributing Red Cross parcels for our barracks unit and took part in the trips down to the tiny railroad station to unload food trains and bring the parcels back to camp by wagon. No horses were available, so the large wagon was pulled by twelve men on either side of the long wagon tongue. It was hard work, but they were a lively bunch that enjoyed the exercise and getting out of the camp on parole to see the sights.

I became the clothing supply officer. We received very few clothing items, mostly used, to issue or distribute other than the rare shoes and sox, but even that had to be recorded on German Air Force forms in German. I gradually learned the German names used for items on their forms.

Near war's end Hitler called-up all able-bodied men, regardless of age, to form a last ditch army called the *Volksturm*, or People's Army. Uniforms could not be made quickly enough to fill this sudden, huge need because of the time element and shortages of material. Instead he ordered that each soldier on active duty, but not serving in the field, would give up his second uniform to outfit the new men. As a supply officer I received the same order, in German of course, so I designated almost everyone, especially those who worked on the ice rink and other sports locations as outside workers who needed both uniforms, even though we only had one apiece. I turned the paperwork over to the German supply feldwebel (sergeant) and never heard any more about it. Sounded like the *Mickey-Mouse* orders we sometimes received then and later. The clerk who wrote this order in Berlin must later have been a part of our U S Air Force Strategic Air Command supply system.

As the time lengthened, I was able to trade for American clothing items for my own use. My German issued military hob-nailed high shoes were later replaced with used but good American GI-shoes. I cleaned them with brown soap "down to the bone" with only the leather left, then polished them like glass. The shoe polish also came from the trader. Many in the barracks thought that as supply officer I had issued new shoes to myself, since the shoes looked so good. By the time the war ended I'd traded for a dark green officer's shirt, pink elastic trousers, plus belt and khaki tie. Never did find an overseas cap though. This was a far cry from the ridiculous blue bunny heated-suit with which I'd parachuted into the snow on the hillside near Angerbach many months earlier.

News

Our news, though somewhat limited, was generally up-to-date. People in the States were told that Hitler kept his people in the dark about what was happening, but we found as time went by that this wasn't the case. From the German side we received the local paper, the Stralsunder Zeitung (Press) from the fair-sized nearby town to the east, on an almost regular basis. The papers gave many more details in most cases than we received from Allied censors. We also got The Berlin Illustrated, a Sunday paper, and the Hamburger Zeitung, on most days, even though they usually were a day late. Once I saw the Rostock paper from the west, but I gather that neither the German NCOs or officers read it regularly. Various camp officials dropped off their papers after they'd read them when they came to the compound for roll call or other business. We passed them around until they were bits and pieces, because there were always enough German speaker among us to read them aloud.

We also regularly received copies of the German Air Force monthly magazine, *Der Adler* (*The Eagle*). Editions were printed not only in German, but also English, Italian, and French for worldwide distribution - as was the case with the German Army magazine, *Signal*.

We had our own resident foreign correspondent, Lowell Bennett of *INS* (International News Service). We called him "Ribbons" Bennett, because he wore a U S army officers green blouse with three rows of ribbons, a correspondent's shoulder patch, green trousers, and another similar patch on an overseas cap.

In the early days of WWII he'd driven an ambulance with the American Volunteer Group during the short war in France. Here he was an internee rather than a POW. After France was overrun he disappeared into the landscape and began writing news releases from the various capitals of the occupied nations. The Gestapo couldn't catch him as he flitted around with the probable help of former local newsmen. After quite a chase the red-faced Gestapo stumbled over his lair and he wound up at Barth-on-the-Baltic with us, though I don't remember in what city he was finally found - probably Berlin.

Late in the war Lowell posted a notice on our barracks bulletin board which everyone came to read. It was at the time when the Canadian airborne were pounding eastward toward us from the West and the Russians were driving across Poland. It said, in so many words, that Allied forces of about ten thousand men had made airborne landings at latitude and longitude so-and-so. Since we were hoping to be liberated and cut-off from the rest of Germany by such a force we believed the story. When we figured out the location we realized it was a gag. What Bennett had actually said was that Allied forces (us) had made airborne landings (by parachute or crash landings) and journeyed to where we were now.

To congratulate Lowell on his sense of humor, and to show our disappointment, he was invited to our barracks. He was hustled into a latrine stall, the door was held shut, while a bucket brigade poured the local very cold water in a long steady stream until he was too cold even to whimper. Finally the Senior Allied Officer, Col Hub Zemke, came by, also taken in by Lowell's story, and said, "Boys, You might let him out now. He's already turned blue." He was truly a drowned rat, and so frozen he had to be supported on each side by two men. I'm sure our treatment improved his judgement.

Letters And Packages

Officers were given two small fold-over letters, like our wartime "V-mail," and three small postcards when available, to mail out each month. The NCOs got two small postcards. No postage was necessary. This was normal in the early days, but thinned out as the war went on.

The first letter to my father contained a small outline about being captured, but I couldn't mention much in the way of details about flying, where I was captured, or where I was. The letter did hold a list of things I needed when he could send a parcel. We routinely added a few nice phrases to please the German censor. I was able to put in a few coded references that we'd arranged before I left for England. I got his answer back, and a parcel which held all the things I'd asked for - eight months later. And you thought your mail was slow! At least, the price was right - no postage cost.

I received a second parcel from my father before the war ended. Pleasant surprises were the two parcels I also received from the German-born wife of the German-American superintendent of the large heavy industrial plant where I'd been a machinist apprentice before the war. I don't know where she obtained the special labels, usually given only to immediate POW families, for mailing to an enemy nation during wartime, but I was very grateful for them. Families were given three such labels per year, but there were no mail limits.

Defeating The Censor

One early letter went to a young lady in Austin, Texas, I'd known while I was a cadet. She knew instructors at the nearby Advanced Navigation School at San Marcos. In it I mentioned that the Sun rose only twelve degrees and twenty-three minutes at noon on the 21st of December - even though the letter was sent in March. I figured that she'd be puzzled by this reference enough to show it to one of her navigator instructor friends. Being very bright, she did just as I'd hoped. Being a navigator he realized that I'd sent the latitude of fifty-four degrees and ten minutes for our prison camp at Stalag Luft One. No, I didn't have a sextant to measure the Sun's elevation. I used the Sun's southern declination for the 21st of December of 23 degrees and 27 minutes South, and the camp latitude to mathematically compute the needed elevation of the Sun. This individual at the San Marcos, Texas school forwarded the information through channels to Washington. The Pentagon sent my father a telegram saying that they had located our camp. At the same time I sent my father a postcard. In the date line I wrote Barth instead of March. My father sent that to Washington also. The German censors did not catch either of my references. In the first letter I gave the latitude, then with the town name I sent the longitude. It took a lot less time to do it than to tell about it here.

I wish I could have sent the same latitude and longitude to the RAF. One of the British air crews who arrived later told us that they'd been briefed on the Flak School nearby as a secondary target.

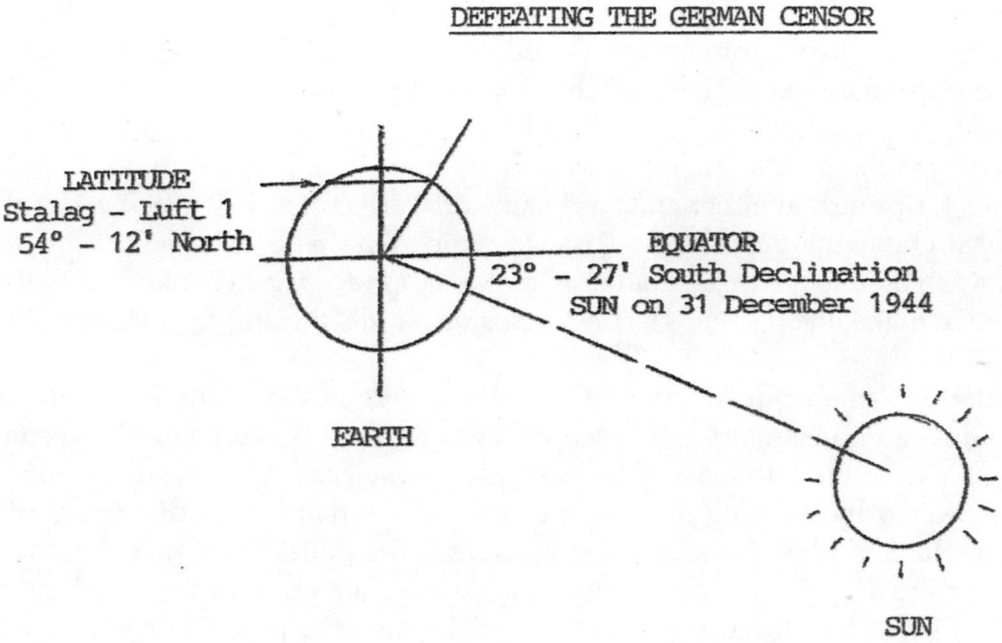

Drawing of arrival at correct latitude and longitude for our Stalag

Library

Over time our compound established a good library in the same barracks #7 where we'd first been quartered before our move to barracks #10. When we moved, a vacant room was set aside for the library. Student groups, schools, colleges, and even governments, plus all sorts of donors, furnished most of the classics and the school books that enabled our many classes to be held in many fields of learning. We had books in many languages on many subjects. I remember seeing one large crate brought into the compound by wagon. It was about four feet on a side and was completely filled with paperbacks of all types - mostly school or non-fiction. It took a whole group of men to lift it from the wagon. One of our roommates, Lt Arthur W Starratt, volunteered with several others, as librarians. Art spent most of his waking hours there. I was able to read many of the classics that I'd never found time to read earlier. He told me that he had helped make shelves for the more than 20,000 books.

Classes

Study classes were held in as many fields as there were instructors and study materials available. Classes were held in any space available. One such class in advanced German was conducted on Sunday afternoons, following the Sunday Catholic Mass in the mess hall led by Father Charlton, the Scottish chaplain. This German class was taught by the older Luftwaffe major in charge of our compound. Even though I was nowhere near the advanced group, I wish I'd found the time to attend every class. I taught my own class in drafting-tool making and mechanical drawing with Ross' art group, and studied other things, but I never got to the language class.

Camp Technology

Like most others, I made the utensils and gadgets I needed, such as a bread toaster made from fence wire, a potato ricer from bed-slat wood and an ever-present Klim can, a Kriegie lamp made from a jelly can, an ice-cream freezer and a washing machine from what else - a Klim can and a bucket, shelves from wire and wood, picture frames, plus models of planes, and many, many, more I've probably forgotten. No, we didn't invent all these things, we merely learned from those who arrived earlier. The RAF chaps weren't always as inventive as the Yanks, but they had started much earlier.

My best effort was the treadle-operated wood lathe mounted between the side boards of my bunk with nails for centers at either end, and shoe-strings as drive belts. It was mostly used for turning model aircraft engine cowlings. Glass was broken to shape the lathe cutting tools. Finely ground glass was glued to heavy box paper to make sandpaper. You shake it onto the glue like salt or pepper. To make it you grind glass against stone and sift it through loose cloth for a very fine finish.

My original glue came from scraping off the small glue beads at the joints of our wooden benches and tables and then melting it in hot water to pick out the tiny wood chips and make it usable. Next, I learned how to make glue from used x-ray film stolen from the medical clinic. The film was cleaned by scraping off the emulsion. It was then finely diced and dissolved in acetone which you could buy for a few cigarettes from the Polish male medical orderlies. This glue was also waterproof. The British learned to make casein glue from the powdered milk Klim that came in the Red Cross food parcel, but I never learned the process or what its limitations were. Corrugated paper from the large cartons that held four eleven pound Red Cross parcels was steamed apart for the heavy cardboard inner and outer layers. The inside corrugation was hung to dry, then sawed over a table edge to remove the sharp water-glue particles to make a very smooth toilet paper and saved to use during times of shortage.

The medical orderlies at the clinic in West Compound had several side businesses. One was in casting soft, low-heat, metal objects in the plaster of Paris. The tooth paste tubes of that era were made from rolled lead sheet, not the thin plastic used now. When emptied, the tubes were collected, washed out well, and the lead melted at a relatively low temperature. Molds of the objects wanted, such as wings and other emblems, were cast from plaster used to encase broken bones that when dried was filled with melted lead. The finished products were sold for the cigarette currency by the Polish male nurses who didn't receive cigarettes as a ration. Being very bright they did good work at a profit.

I still have my *kriegie* "wings". (Please see Color Plate 1 for illustration) They were cast with lead studs on the back. With bored holes and a piece of wire they were attached to the uniform in lieu of a pin back. This design started with American pilot wings, but with the inset vertical bars removed from the shield. In that place were the letters POW. Above the shield is the representation of a parachute, since that is the way most of us arrived in Germany.

Not only were American wings reproduced, but pilot badges of all the air forces in Europe were copied from the actual badges. In a raffle, naturally for cigarette coinage, I won a photo album whose front cover held eight pilot badges, including a German one. It was one of the things stolen from my tent at the RAMP (Recovered Allied Military Prisoners) camp Lucky Strike near Fecamp, France, on the Channel coast. When we went to the mess tent to eat, many of our tents were raided by the non-combat stevedore troops who stole anything of value they could find. Welcome home.

Entertainments

Camp entertainment came in many forms. On Sundays and holidays, the *Round The Benders* orchestra with Mike Spodar, the trumpet player mentioned earlier, lead by musical director Lt Harry Korger, played during meals. Harry Korger was also a good song writer. I remember some of his songs and orchestrations. Harry was also one of the lead navigators on the famous, but-ill-fated, Ploesti oil refinery raid in Romania that cost so many lives.

The orchestra used band instruments marked "US". Again, our country donated the musical instruments to neutral Switzerland, who in turn donated them and shipped them to our camp in Germany.

One very unusual thing was that Mike Spodar, during the Winter of 1944-1945, was permitted "safe conduct" to leave his barracks in the very cold evenings to play his beautiful music across the snow to us. It was called *Silver Trumpet*. He played nostalgic songs and classics, both American and German. I'm sure the guards loved it as much as we did. Anyone else would have been fired at if seen outside of their barracks at night without permission.

On some occasions, usually at the evening meal on Sundays, we had "stand-up" routines in front of make-believe microphones by a group who pretended to be broadcasting the comic radio shows then popular in the US. One such was a wild and racy version of the radio serial, *Ma Perkins*. Some pretended to be various well known radio comedians. They did very well. Of course, we were an 'easy' audience. We enjoyed all that was presented.

Three Jewish-American lieutenants, Levy, Feldman, and Gerber did vaudeville or burlesque routines in various accents. Many were really hilarious. One gave readings and acted out from library books in a heavy Yiddish accent. One I particularly remember very well was the tongue-tied version of Shakespeare's Soliloquy. It was given by one of the three, but I don't remember which. There was a great amount of other talent on hand, too.

Near the war's end, when an order came from Berlin that Jewish prisoners would be segregated in a special barracks, # 11, these three men told the Germans that the "H" on their dog-tags meant Hindu, *not* Hebrew. The three comedians made turbans and prayer rugs to carry out their gag. The Germans, who had to respond to the Berlin order, kept a straight face and made up certificates for each to sign. The papers were sent off to Berlin. Nothing ever came back. Sounds like the Pentagon.

One crazy, whose name I've forgotten, came out from the kitchen to the mess hall podium in a brown overcoat, black hair and Hitler mustache, wearing an imitation Nazi armband. Although he didn't speak, or even understand, German, he had a gibberish that sounded like Hitler's speeches. He'd start up slowly and reasonably, then grow increasingly loud and wild. His harangue and arm-waving mannerisms, which made no sense, did mimic Hitler very well. On one occasion, when he was in the middle of his loud *spiel*, the main entrance door at the back of the mess hall, our entrance, opened and framed our German compound commander, with his NCO behind him. The major backed up and slammed the door closed. The speaker ducked again into the kitchen. The major re-entered, a few moments later, his face purple, not from anger, but I think, trying to stifle a real belly laugh, from what little he'd seen. He acted as if he'd seen nothing. The NCO was completely blank-faced.

We were shown four American-made *Andy Hardy* movies on bed sheet screens over the Christmas and New Year's holidays of 1944. We were told the copies were bought by the German Air Force in Switzerland. They starred Mickey Rooney and Andrew Stone, plus several young girls then popular.

There were many plays put on by the prisoners themselves. Of those I remember best were: *The Man Who Came To Dinner, Front Page,* and *Dangerous Corners.* These were not sophisticated actors, but we were not a sophisticated audience either. In *The Man Who Came To Dinner,* our roommate, bombardier, tall, lanky Lt Arthur "Herky" W Starratt, played the housekeeper. Instead of staying in the background except when he was supposed to deliver his lines, Herky, as we called him, filled the play with much more slapstick comedy and antics than in the original written version. In one laughable scene he stole, he noisily dropped a pile of metal kitchenware in the middle of Whiteside's big speech. That actor wasn't happy, but the audience's roar drowned out the lead actor.

The funniest slapstick I've ever seen was *Kriegmalion*, played by Lt Alfred Sheppard, from Nashville, as a drunken sculptor; and by Lt Eugene "Sabretooth" Johnson, of Redwood City, California, as the female sculpture being created. It was originally written, I suppose, but I saw it four times and it followed no discernable script, but was wildly ribald and different each time.

In one skit with Sheppard and Johnson, Johnson was a farmer who came to consult with Sheppard who was made up as a mystic with a robe and a turban. The story line was that the grandfather had died, was buried, and no one could find the key to the outhouse. Johnson came to have Sheppard consult the "Great Spirit" to have him contact the deceased grandfather. The funny line that really resonated with the airmen present was, "Oh, Great Spirit, do you hear me? Over."

They were spontaneously funny, though neither had known the other before they were assigned to the same barracks room. "Field Day" on the 4th of July, 1944, among a whole day's events, a ring with ropes was set up for actual boxing and wrestling events. During a break, Sheppard and Sabretooth climbed into the ring and did a comedy routine that lampooned our American wrestling. Not only the American audience doubled over in laughter, but so did the German Air Force attendees, both officer and enlisted.

There were many more similar things too numerous to count or remember. No it wasn't idyllic, no prison camps are, but our *Kriegies* made the most from what was available. Our keepers generally were honorable and professional and allowed us as much freedom as was prudent and practical.

Sports

At our young age, sport was of great interest. We had volleyball, softball, and eventually ice-hockey which began in the late Fall of 1944 with the arrival of clamp-on, key-tightened, skates of very old model. They were donated by student groups from Switzerland. We learned how to keep them on our feet by wrapping the skate to the shoe with strips of wet cloth. When outside in the Baltic winter they froze and held the shoe and skate together very well.

Except for the clamp-on ice skates, we received only hockey sticks and a few pucks. Everything else we had to make. Three pairs of socks over our hands became gloves. There were no pads, goalies simply wore all the clothes they could find. I don't remember anyone losing teeth. The goal nets were made with wooden frames and a netting braided from three strands of the wood fiber called *excelsior* in our country. It was taken from the wood fiber mattresses we slept upon.

In our North One compound we built the needed winter ice-rink for general skating sessions and for hockey competition. The ice-rink was made by clearing the snow from a flat area in the needed dimensions from the open space between barracks #4 and #7. The cleared snow was formed into a small berm that was wet down to form an ice barrier along each side of the rink. Then a bucket brigade brought cold water from a latrine window in barracks #4 to build up the ice surface. It took us two days to bring enough water, because the guards would not loan us a hose.

Those of us who instigated the operation and actually built it, took a permanent pair of skates since we were also the maintenance crew who kept the ice surface repaired after each session or ice hockey game. We didn't know the term then, and didn't have a vehicle, but we became the human *Zamboni* to re-surface the ice after each session or when needed.

It was said that our camp volleyball caused more injuries than the war. Most were caused by slamming down across the net. The ferocious competition caused more collar-bone injuries than had been caused by earlier aircraft crashes and by parachute opening shock.

Softball caused a few collisions, usually with only minor injuries, but more concern was created by some oaf breaking our only bat. Replacements were hard to find. We played at times with a borrowed cricket bat. They never broke, but weren't too satisfactory. With the diamond shaped cross-section and angled hitting surface, the ball either went into the ground or high in the air, unless you were able to hit the ball with the exact center of the bat. For awhile we even used a bat carved from a stolen fence post taken from a warning wire . It gave many splinters, but didn't break. Bases were made from bags filled with sand. If a ball rolled under a *warning wire* near the fences, you had to get permission to retrieve it from the nearest tower guard before reaching under the wire and being fired at.

I played two abbreviated seasons at second base with the *Repats*. This was a short version of the word *repatriates*. Men with mental problems severe enough were sometimes repatriated to the US by ship on the neutral Swedish Red Cross ship called the *Gripsholm*. One of our Catholic priests, a civilian captured in North Africa, went home by this route after he went *around the bend.*

I had a great batting average, and loved it. We laid out two diamonds at either end of our parade ground. We couldn't have played baseball, even if we had the equipment, due to the longer hitting range We played even during rain or snow. Our team names, were often from German newspaper insults against us, such as *Terror-fliegers,* (Terror flyers), *Luft-Huns,* (Air Barbarians), and the like. Many were merely ribald, such as *The Family Jewels*.

The Sound System

An aid group, perhaps the YMCA, in the U S donated a sound system, a professional record player, speakers for the barracks, and musical records for our use. Its installation was held up for several months, because the German camp command wanted the new electrical wire for their uses, and wanted to give us used wire. Our leaders didn't want to help the German war effort by providing them with the new wire, so there was an impasse for awhile.

Finally an agreement, or compromise, was reached. The Germans would get the new wire, which they could have taken anyway, and we'd get used wire. Under the agreement the Germans would wire the building speakers, and would provide an English speaking German NCO to be our DJ and play records for us for one hour in each wired barracks after our evening meal. The NCO would also read to us the song names and whatever other information he chose about the music. In addition the German command would also use this system to provide information and orders to us

This is where we first heard Spike Jones and his raucous, *Tea For Two*. The German sergeant must have played this several times for himself to get the timing down, before he played it for us, because he turned up the volume very high just when the music went from slow and gentle, to the very loud screeching and banging. I was sitting with my chair leaning back steeply against the wall. At the uproar my seat slid out from under me, and I fell to the floor, while wondering what had happened. I'd like to see a picture now of what our faces looked like then. From then until war's end we loved the music.

Payroll

Under the terms of the Geneva Convention of 1929, which both the U S and Germany signed, we each were paid a specific amount every month, depending on officer rank, in Kriegsmarks, or war money. It could only be spent at camp. It was like the Post Exchange *chits* we received while we were aviation cadets. American officers were not paid individually, their money was paid to our paymaster. This was good, since nothing was later deducted from the pay accrued while they were prisoners. RAF officers did receive their pay directly, and the equivalent amount was deducted from their back-pay when they returned to England or Commonwealth countries.

Our paymaster, Captain "Dixie" Lee, a Finance Officer with a bomber group from the 15th Air Force in Italy, managed to get shot down and captured on an easy *milk run* mission. He'd always wanted to fly a combat mission. He got his wish, but was shot down - and we got a finance officer who knew all the pay rules. He lived in barracks #9 with the senior officers and was available to answer all our questions about pay and allowances. It gave us another thing to look forward to.

He also maintained the records as to our German pay - and he handled purchases where our pooled fund was permitted to buy extra razor blades over and above the one per month provided by the Stalag, and anything else that was needed and available, but not supplied by the German command. Our Senior Officer, Col Zemke, required that all officers and NCO's share equally in whatever could be bought, regardless of rank, since NCO's did not receive pay. Dixie Lee was also able to use our joint "income" to buy small necessities or extras such as tooth powder, sewing materials, items for the art and other schools, etc.

Although this seems very improbable or maybe hilarious, some of this money was deducted by the German authorities to pay for the wire we stole from the barbed wire fences. During air raid "blackouts" we went out unseen and cut wire from the fences, then came back into the barracks to removed the barbs with our special tool. The wire was cut by two ice-skate blades riveted together, using long wooden handles for leverage. We would have given back the barbs if there had been a rebate or discount. I don't know if anyone ever found a use for them.

Another major deduction from our pooled money came about when the Luftwaffe made us pay for all the stolen wood and extracted nails from the barracks building. If the German carpenters hadn't come around on a regular basis, the buildings would have collapsed. For repairs usually we'd see an older German NCO with his bow-saw, followed by two Russian helpers carrying tools and a few boards.

The Parole System

A parole system was in place that allowed the Senior Allied Officer to put on a purple armband with a white cross and walk out of his barracks, day or night, out through the double gate, and go to the German headquarters for needed consultation. This system was also used by Father Charlton, the Scottish Chaplain, to go wherever he was needed. In many cases, I'm told, he walked into the small town of Barth nearby to have a glass of beer with townsmen friends at the beerstube where he was apparently well known, liked and respected. All other outside groups had to be escorted, such as volunteer work parties of young officers. NCOs could be required only to work as supervisors. Officers were not required to work, but we all took turns doing volunteer work whenever we could help. We even did KP. Of course, we did our own housekeeping, since maid service was not provided.

As mentioned previously, officers were allowed out on parole for recreational purposes for limited time periods and were guided by an English-speaking, unarmed, German NCO. We signed the parole roster for each trip and were escorted in groups of 200 or so to hike out to the waters of the Boddensee on the west side of our peninsula for swimming in the Summer. I wish I'd been able to fish, but I did use these trips to gather black mussels, which were very plentifully clustered on reed stems in the shallow water. I brought them back in a bucket to steam for dinner. On other parole trips we had hikes northward on the peninsula where our camp was located. We stayed clear of the wooded areas north of the camp location that were used for German defensive field exercises and the open areas used for the training of the Luftwaffe Flak School students. The trainees and their instructors used vehicles to tow the large Wurzburg antiaircraft radar units and the Flak 18, 88mm cannons on the road past our barracks to and from their field areas.

As an aside note, the Flak School on our military reservation and just south of our compounds took on much greater importance and virtually all of the gunnery and radar training then being done in Germany after the former main flak school near Munich was destroyed earlier by Royal Air Force bombers. The Flak School buildings are used today as the local high school.

Allied NCO's, under the terms of the Geneva Convention of 1929, were not authorized to sign-out for parole. However, any officer was permitted to bracket the names of one or more NCO's with his name and accept responsibility for their return.

In one rather funny note, that not only shows the humor and devilment of the American detainees, but also their vast acquisitiveness, one collected item stands far out. Firewood could only be collected from the ground after it had fallen from a tree. In the Fall of 1944 an officer party on a parole hike north of the camp was allowed to collect downed tree limbs to bring back with them for firewood. While doing that, one individual found a complete, live 88mm, *Sprëngranate* (High-explosive) Flak-18 round that must have been lost in the tall grass by the German Air Force students during loading practice. In any case he picked it up, hid it under a group of tree branches, and smuggled it back with group to his room in barracks #1. It's difficult to see how he got it past the escort and the gate guards, since the complete shell, with case and projectile, is more than 36 ½" long. He also must have been very strong, because that shell, fully loaded, weighed about 40 pounds. This is the ammunition that shot down so many of our B-17s and B-24s.

The *kriegies* polished the shell case to a mirror finish and used it as a centerpiece for the table in their room. A Luftwaffe captain discovered it upon entering their room sometime later. Can you even imagine his shock, surprise, and wonder? No, it wasn't like the TV comedy, but it must have been very close to that.

The Jewish Barracks

In the Fall of 1944 an order was received from Hermann Goering's Luftwaffe headquarters in Berlin and posted in German on our bulletin boards. It said in effect that since German airmen were not required to live with Jews, Allied airmen would no longer be required to live with Jews either.

The prison camp command decided that a former warehouse to the rear of our barracks, #10, and separated from us by the normal 10 feet high and 10 feet wide barbed wire fence, would become the required home of all Allied Jewish prisoners. They would be moved from the other compounds as well as our now North One Compound. A new fence was built on the outside of barracks #11, and a gateway was cut through the fence section facing barracks #8. The open building was remodeled and rebuilt with individual rooms.

The Jews arrived, moved in, and lived normally. They feared that they were being segregated and would be sent to concentration camps. Nothing bad happened to them. Their lives remained the same in their new barracks; they continued to eat the same meals along with us at our mess until that building later caught fire and burned to the ground. They were liberated as we were when the German command turned the camp over to Allied control.

The American Training Planes

On a nice afternoon during the Summer of 1944, while sunbathing on a blanket, I heard a familiar sound. It was the drone of a small formation of our 2-seat, low-wing, single-engine basic flight trainers. When I looked up I could see a group of five BT-9 Fleetwing basic trainers practicing formation flying. They were very easy to recognize because of the fairing that covered the fixed landing gear struts. We called them *pants*. In the U S Army Air Forces, cadets began their training in a primary trainer such as the Boeing Stearman biplane, then went on at the next level to a basic trainer such as this, before finishing up advanced training in a single engine AT-6 Texan or various models of twin-engine trainers.

I jumped up, yelled, and tried to tell men nearby the plane's name, but only got back comments such as "you're nuts! They're probably Arado 56's, or something like that, with German cadets." No one would believe me.

That evening after roll-call I caught up with the German major after we were dismissed - and asked him where the *Luftwaffe* got the American Bt-9's. He grinned and said, "Oh, you noticed. We took them when France fell. He said they use too much fuel and you won't give us spare parts so they'll soon be gone," and walked away.

Casualties

During my stay at Stalag-Luft 1 there were several deaths, and at least several severe woundings. One man died from encephalitis, possibly mosquito-borne, due to our surrounding waterways and marshes.

Another man in a nearby barracks persisted in hanging laundry out of a barracks window during an air raid when we were required to be indoors. The tower guard fired his rifle and the bullet passed through the lieutenant's back and out through the chest, missing the spine and major organs. After emergency care and hospitalization at a Luftwaffe hospital he recovered fully.

One young officer woke from a morning nap, not realizing that the air raid warning had been sounded. He got up and headed from his barracks across the compound toward the Sunday morning mass at the mess hall. When he realized that there was no one around he wheeled around, knew something was wrong, and headed back. A tower guard spotted him, fired, and he died instantly.

A young man in our barracks had his skull fractured by a very old guard who swung his rifle in fear. At the end of *Appell*, when the group was dismissed, a small mob pressed through the fence opening to our barracks area. The old guard who covered this opening during roll call did not move away soon enough and thought that the group of boisterous young men were going to attack him. He swung the rifle butt wildly. Most everyone had moved clear except for the man who wasn't looking in that direction.

There were similar incidents in the other compounds, caused by similar circumstances. The normal *Luftwaffe* NCO's and guards were no danger to us, but the untrained, easily startled and

fearful old men were a real danger due to their lack of military inexperience.
V2-Rocket Launch

One evening just after sundown during *Appell*, while we stood At Ease, we saw a thin white line climbing to the northeast. It rose into the reddish twilight then disappeared. A bit later a pink ball appeared. We could see that it was a rocket launch, although we knew nothing of the German rocket program. Somewhat later we heard an explosion and knew what the "ball" had been. When our staff asked what it was, they were told by the German compound commander that it was a high-altitude weather rocket from the nearby Stralsund Naval Air Station. I don't know whether he knew better, or whether he'd been told the same thing.

Col Greening included it in one of his watercolor sketches in his book, *Not As Briefed,* published after WWII.

In fact it was a German A4 rocket test from the research base at Peenemunde, east of us, and fired out over the Baltic Sea. It was one of Hitler's *Vergeltungswaffen* (Revenge or Retribution Weapons) which arrived too late to stem the tide of our retribution. It's where the *V* comes in V1 & V2.

The Flintstone Flak Battery

This was of a lighter note. No, we hadn't heard of the Flintstones back then. I added that name much more recently, because it is so appropriate. The German flying cadets received their advanced training at the nearby *Luftwaffe* air base at Barther Fliegerhorst, the neaby air base. They, much like their American counterparts, loved unauthorized activities such as "buzzing" people and places on the ground in their single engine Arado trainers when there was no instructor aboard

As the opportunity arose while flying solo, the cadets would dive very low into our compound's parade ground area, then pull up to barely clear the barracks at the far end. We liked seeing them, but as befitted our very young age, we decided to give one of them a *dusting* or scare.

The first group of five or more recruited several times that many. Each selected a round stone about the size of a plum which could be hurled upward as the cadet's plane passed overhead. A leader without a stone was selected as battery commander to face the approaching plane and give the "fire" command as we faced away, so we could put up a proper, dense barrage.

On one nice Spring afternoon a cadet zoomed in. By the time he went around for a second pass we had time to stop whatever we were doing and assemble with our "ammunition". On approach he must have seen our group of 20 or 25 forming and decided to give us a real "dusting". He dove in, engine howling and as he passed overhead we fired on command. No, we didn't aim, we just threw upward as he flew by. Even from the ground we heard the clatter and many bangs. I imagine that the cadet was startled by our reception - and that he was really "chewed-out" when he landed, because I was told by a German NCO, later, that the plane had propeller damage, a cracked canopy, and dents all over the wings leading edge and fuselage.

We were told the prison camp staff was very angry, and, "You do dot again, und ve vill shoot."

No cadets came back, the game ended, and we had to be content with our only "victory".

Real Flak

We were told by our long-timer RAF friends that years before the Yanks arrived at Barth, a guard tower was shot to pieces by a low-flying RAF Lancaster bomber limping out to the Baltic Sea on three engines after a night attack against Berlin. In the pre-dawn light, a German tower guard with more courage than sense opened fire with his twin, light, aircraft-type, machine guns at the bomber roaring by *on the deck* nearby.

The bomber crew probably hadn't even seen the blacked-out camp, but they did see the bright gun flashes. This "rudeness" or impudence really *browned* them off, because they came around in a sweeping turn with all four front-turret guns blazing to *return the favor*. The tower was blown apart and the inexperienced guard was *hamburgered*. After that, I'm told, the guards were warned not to do that again. Evidently, in this case, discretion is the better part of valor.

There are, no doubt, as many actual anecdotes as there were prisoners. Now they are mostly long forgotten and buried with the airmen who spent a few, or many, days, weeks, months, or even years at Stalag-Luft One. In my case it was one of the most interesting periods of my life.

12
Escape Attempts

The first duty of a prisoner of war is to escape from captivity and return to his own country's forces. For us this held true until we received verbal orders that we were to remain where we were until we were liberated by Allied forces. This order from General Eisenhower's SHAEF Headquarters after the cross Channel invasion of June 6, 1944, and other special instructions to POW's, were orally transmitted to the Senior Allied Officer (SAO) at the various officer prison camps. This was done by senior officers who themselves were shot down and captured while flying as air combat commanders or senior observers. Before their flights these officers received the latest orders from their higher headquarters for use in case of capture.

I'm sure that a similar method of transmitting orders to the NCO prison camps was in effect, but I've never heard it mentioned or explained. The senior NCO in such a camp was elected to his position by the other NCO's present and was called "The Man Of Confidence". Date of rank or highest rank was not always used as with officers.

When an individual or group worked out a seemingly possible escape plan, contact was made through the barracks commander with a *runner* from the escape committee. For security purposes no direct contact could be made with this committee. We didn't know the members in any case. A runner was known only to a barracks commander who knew no more then that, but who could steer you to him. This *cut-out* prevented the German security people from *rolling-up* our escape and security groups and knowing whom they should watch most closely.

If you had a plan you could discuss it with your barracks, or *block*, commander who did not have a "yes" or "no" function. The commander was obligated to put you in contact with an *escape runner*. If you didn't wish to discuss your plan for any reason, security being one, with that commander, that was your option. He still made the contact for you.

Next you went for a walk with this individual, singly, and discussed your ideas. You didn't find him - he found you. If more than one POW was to be involved in the plan, the group had to select a spokesman to *front* the plan. At the end of the discussion you had to forget the runner's identity. You didn't get a yes or no answer. If the individual did not get back to you with a request for more information, or give you tentative approval by passing you on to another POW *higher up the chain*, you had your answer. As with many such things, we learned about these arrangements from those who came before us. In our case it was the long-term Royal Air Force prisoners, who had over the years of captivity before our arrival, learned by *tried and true* methods.

To give an actual example, Lt George Kilduff, a roommate and fellow navigator, talked over the idea with me of attempting to escape by pretending to be boys from the surrounding area, We both looked young enough. It was thought that the local police did not often check papers of the young - and that all we'd need was a wave and a greeting in German to safely bypass police officers or their checkpoints. Back then, dressed appropriately, we believed we could still pass as boys. Being small and very thin was a definite asset for that image. Even tall or *gawky* probably helped.

The process would begin with a climb over the selected fence during an air raid blackout when all the lights were off, but couldn't be attempted in mid-to-late Summer. Then it never became dark enough during that season. The very long sunny days were followed by a long luminous twilight before a long and early dawn. There would also be many more people out and about, walking and enjoying the fine weather. We'd have to stay away even from civilians, because we'd be noticed when we couldn't understand their questions or answer properly in German.

Long benches, made from wood stolen from the barracks would span the ten foot distance between the inner and outer barbed fence wire tops and get us quickly past the rolls of 'concertina' wire in-between. This method would require one or more men remaining behind to remove this 'bridge' after the passage. Since the plan was for the very late Spring or early Summer when the mornings were still cool, our clothing made by volunteers with tailoring backgrounds would consist of heavy cloth shorts in the style of *lederhosen*, short-sleeved shirts with colorful embroidery, RAF white *submarine* sweaters with the long arms tied around the neck, high socks of any kind as long as they didn't match, and no hat. All the adults wore hats, but teens didn't. We'd carry books fastened by a strap, walk only in daylight, though never at night which earned the most inspection. We'd stay in the open during the day and sleep in hedges. Carrying a small sack for food supplies over the shoulder would be normal for boys, too.

We'd head eastward to the nearby seaport at the town of Stralsund. If no boats with Swedish flags were in we would continue down the coast farther to Stettin. Though it's now across the border and has a Polish name, it was a bigger port then and usually had many ships in its harbor. There, coal and ore boats arrived from neutral Sweden with raw materials that were vitally needed for Germany's war effort. Germany still has only low-grade coal and poor quality iron ore.

The larger ships displayed various signals for which we would have been briefed before departure. The signals changed from time to time. These markers would let us know whether or not we'd be welcomed aboard. The escape committee was somehow informed of the latest rules. I guess the information was transmitted as needed by coded messages in the BBC programs received on our camp's hidden radios. One such signal I remember was a pair of *long-johns* drying on the ship rail to the left of the gangway. If it was there you could join the loading gang of stevedores carrying supplies aboard and then be taken below to the crew's quarters and hidden in a bunk until the ship left Germany's territorial waters in the Baltic Sea. Supposedly the ships and dock areas were not carefully checked on a regular basis for outbound stowaways. On the surface, Sweden was a neutral nation, but in fact its people for the most part favored the Allies and did many things to tilt toward the West. Germany also needed neutrals like Sweden and Switzerland as conduits to the world.

George and I cut cards to see who'd meet the runner. I won. A day or so after we spoke to the barracks commander, I was approached by a fellow I didn't know - and from a different barracks. He said, "Hi, Joe and George." I knew what he meant. We discussed the escape idea as we walked around the compound's outer fence. He asked a few questions that were answered by the things you read here, then walked away.

The next day while we were setting up for the drafting class I taught, Lt Col Ross Greening, the Art Committee's leader, came nearby and with a nod motioned to me to walk away from my group. Ross then very quietly told me that I couldn't leave and was needed where I was. Afterwards I told George Kilduff the bad news. Even though we were roommates, I never knew if he teamed up with anyone else to follow up on the plan. He wouldn't have told me, in any case, if something else was in the works.

The "stay where you are" order from SHAEF didn't come to us until much later, after the Normandy Invasion. I was sorry we didn't get the chance to try our luck. After all, we were only about 60 miles across the water from Malmo, Sweden. We'd have loved to make the attempt and perhaps visit Sweden after a Scandinavian cruise, even if it was only on a coal or ore boat. It would have been a fitting *follow-up* to the long train excursion we'd had from the *Luftwaffe* interrogation center at Oberusal in the small boxcar whose floor under the straw was covered in very finely ground coal briquet dust. I never saw a mirror when we arrived at Barth or knew if we were black enough for a *minstrel show*.

We heard that the Swedish internees had a pretty good life and were even allowed to visit other parts of Sweden while *on leave*. Also, many of the Americans who landed there, in contravention of neutrality agreements, were illegally smuggled back to England for return to the US, along with escapees coming across the border from German-occupied Norway to wait out the war.

It was also humorously mentioned that the many German fighter pilot internees there weren't very eager to leave - at least not while the war was still going on. Another joke was that you had to be very careful when landing at a Swedish airbase so you wouldn't be hit by one of the many German fighter planes in the landing traffic pattern.

The Americans who managed to fly their battle-damaged planes to safety in Switzerland, however, were made to stay in camp until the war ended, even if not actually under confinement as prisoners.

Ross Greening, also the North One compound commander, and Commanding Officer of Group One, roomed in barracks #9 with Lt Col Loren G McCollom, the camp's intelligence officer, on the staff of the Senior Allied Officer, then Colonel Jean R Byerly. It wasn't too difficult to figure that our escape plan had gotten to the top very quickly and had been just as quickly *shot down*. I don't think the plan lacked feasibility, but I never heard of it being tried in our or any other compound.

The art class, unknown to most of its students, and virtually everyone else in North One compound, was also the cover for the Map & Document section of the Escape Committee. We copied escape maps, forged gate passes, and made whatever minor documents or drawings needed. The secret items were copied, drawn, or worked upon by only four of the instructors. The balance of actual students were our *cover*. As our British *lads* would have said, "it was fun sticking our thumb in Hitler's eye."'

Our map information source was the cloth maps of various areas of Europe that had been sewn into the linings of uniforms jackets of the flyers of the British Royal Air Force who were now prisoners. The RAF fellows in the West Compound made copies of these cloth map sections and smuggled them to our compound for us to copy. Our group then made copies of copies as needed.

This became the biggest, by volume, of our "drafting" jobs. We began by slowly and carefully rubbing small amounts of our German supplied mineral oil margarine into sheets of typing paper for water-proofing - and to make the sheets semi-transparent. Ground chalk then was rubbed in to remove excess oil. After I learned how to do this from the RAF, I trained two more men to help.

In addition to the cloth maps, some of the RAF uniform buttons also contained very small compasses that could give general directions to an escapee when the Sun wasn't visible. The small RAF uniform with which I had been outfitted at the Transient Camp at Oberusal did not contain either map or compass. Yes, I checked as soon as I learned about this.

The US didn't wise up to this map system until very late in WWII. After the war ended I was told that Monopoly games were fitted with both maps and compasses and sent through Switzerland, supposedly to the prison camps. Unfortunately, we never received any, since I'd have been one of the first to know of their arrival when they came to us to be copied.

Another of our jobs was to make simulated gate passes by imitating "typing" with a dip pen. Trial and error and lots of practice with many spoiled copies made it barely possible. It worked quite well if you didn't look too closely at these copies and if they were not used too often in the same location. I don't recall hearing that any of our passes were spotted by the gate guards or guard NCO's. If they had been spotted, we'd certainly have heard about it. To get paper similar to and thick enough for the real item, we had to cut back pages from books in our library. From an actual gate pass I copied the German signature, then practiced until I could sign as many as needed. One of my jokes was that I could sign the German *kommandant's* signature better than he could.

Such passes were used successfully to make unauthorized trips to other compounds to carry needed information and useful contraband. Long before, as a boy, I'd done this imitation type-writing by pen for fun and to fool friends, since I wasn't allowed to use my father's typewriter. My Scottish stepmother once gave me a *clout* before seeing that my printing wasn't actually typing.

Many widely varied things were needed for escape attempts, but it seems that there were always people who had the talent to produce the needed items. In fact, for one plan a *blue* suit was tailored from RAF uniform cloth to enable an individual to join an international Red Cross inspection team during their semi-annual tour of our *stalag*. Unfortunately, the sharp-eyed German main gate guard who had admitted the original party noticed that the group leaving was one man greater than the number who'd earlier arrived - according to the number on his clipboard. He patiently checked the passport identity, one by one, of each member of the visitors team and soon found the added British officer. That chap then got an all-expense-paid trip to the klink. *C'est la vie.* (That's life.) Nothing ventured, nothing gained.

I'm sure that the civilian Swedish Red Cross team members were very well aware that an escapee had joined their ranks just prior to departure. Their required strict neutrality had prevented the inspection team members from helping either the detainee or the Germans. They also couldn't expose the man or help him in any way.

Another very bold escape plan was approved for the captain, who at that time was our barracks commander. He had moved with us when we were moved from our original location in barrack #7 in the middle of the compound to our new home in barracks #10 along and paralleling the western fence.

In essence the plan was to escape from our compound with three other men without being missed long enough overnight, between roll calls, to make it to the nearby operational Luftwaffe airfield, the Barther Fliegerhorst. There they were to steal a Luftwaffe twin-engine aircraft, take-off, and fly across the Baltic Sea to Sweden. While that can be written in just a few sentences, all of the needed elements that were required to come together and make it work would fill the many pages of a small book. The detailed planning, the study of the many elements, and the actual practice took several very full weeks. I'll get back to that shortly after I describe the physical layout the escape team of four would have to pass over and through.

Now that we lived in barracks #10, a former German flying cadet barracks that had recently been converted to house American officer prisoners, we now lived in an area that had originally been outside of North One compound's west fence and along the main north-south road that crossed the camp from south to north. This concrete road still exists. I've even walked down it with wife, son, and friends. Our returning groups have walked along it also each time we've returned to Barth.

The smaller barracks #9 that housed the Senior Allied Officer and his staff plus the Compound commander's staff, and our barracks #10 had earlier been fenced in before our move. An opening also was cut through the earlier compound fence, to make our new area part of North One compound. This gave us access to the mess hall and to the twice daily roll call. However, the next building north of our #10 and separated by a fence, was the still unfenced, barracks-sized, building #11. This building was then only an unguarded warehouse with ready access to the road. Soon after we moved to #10 we also learned that the fence that separated barracks #10 from the unfenced warehouse #11 was visible to only one tower guard, who not only was at a long distance from this fence, but at the extreme southeast corner of North One compound.

An escape plan was soon set up using this near *blind-spot*. After evening roll call, but before dark, a four man aircrew could cross this fence between #10 and #11 rapidly while the one far-away tower guard was looking in a different direction. The crew could take cover in warehouse #11 until well after dark before making their way south to the airfield. On route they would circle the camp clockwise around the northern perimeter fences and come southward through the open potato fields on the east side. That route then was still clear, because construction of

North Two compound had not yet begun. If this almost open route was to be used it had to be exploited before German security awoke to the fact of the blind spot.

To make this daylight fence-crossing fast and almost foolproof, the crew members practiced at great length on a simulated fence top. It was constructed from two stolen heavy fence wires with barbs remaining, as on a real fence, and strung tightly from hand-made hooks attached to the central corridor walls of barracks #10. The wires spanned the corridor almost as high as the fence tops and about the same distance apart as tightly as possible. If the Germans saw these wires in place before they could be removed after practice, they were to be told that the wires were for indoor laundry drying.

For the practice here and for later use at the actual fence crossing, two men not going on the escape came to the wire before the escapees. The first and bigger man would bend over at the wire and a second smaller man would leap upon his back. The upper man would reach back to grab the bridge board handed to him by one of the 'escapers' to cover the gap between and to go over the top fence wires. The crossing board had braces at either end to lock across the wires and keep the board from slipping or sliding. As soon as the board was in place, the captain leading the escape would climb the backs of the two previous men, cross the plank and jump from the far end to tumble and roll on landing on the warehouse side of the fence. Each of the four crewmen had to rapidly do this in turn until all were across the wire while the visible tower guard looked in a different direction. This was practiced endlessly in our hall until each of the two helpers and four aircrewmen could complete the exercise without error in the least possible time. This system could only work in the 'blind spot' between barracks where there was no warning wire that kept prisoners from approaching the main, or outer, fence.

The next training requirement was to attend the flying course so the two pilots could learn how to start the engines, locate and release the parking brakes, taxi the plane to the runway, and then take-off in the captured plane. Bob Hoover, the ground instructor-trainer for this mission, was the stalag's most knowledgeable aviator about Axis aircraft. While in North Africa, Bob had flown quite a few captured or abandoned German and Italian aircraft and knew a great deal about them. As the draftsman, at his direction, I drew sketches of cockpit and instrument panel layouts for crew study. Bob knew generally what and where the instruments for flight and engine controls were located on each plane - and what the words on the instruments meant. Bob also covered engine starting procedures and everything that would get the crew underway. They even discussed how to drop the aircraft's bombs safely in the Baltic Sea.

From the airfield to our south we could see aircraft tails passing from our camp as they taxied by in the distance. Bob determined what planes would probably be operationally available and which ones to study for the program. He covered Dornier and Junker twin-engine light bombers that were in daily use against the Russian Front rapidly approaching from the East. They were rapidly over-running the German forces even at that early date. Normally there would be Dornier DO-17 or 217, plus Junker JU-88 Schnellbombers. If there were other similar aircraft types, they'd probably operate in a likely fashion.

At the airfield our escape crew would head toward and hide near a plane showing activity. Usually the German crew chief and helpers would check out the plane, run-up the engines, then shut down and stand-by for the fuel truck to top off the gas tanks. After fueling, and after the armament crew loaded the bombs and defensive machine guns, a truck would carry the ground crew back to the flight line mess for breakfast. It would be much like the routines we used in our air force. Later the ground crew would come out again in the truck with the flight crew to prepare for take-off.

With possible pursuit in mind after the plane became airborne the American crew planned to salvo the bomb into the Baltic Sea while flying northward to Sweden. This would make the plane lighter and probably somewhat faster. For navigation the crew was furnished with a crude pencil-drawn map that would not alert their intelligence to our map-making ability. We didn't want the attention.

After weeks of fence crossing practice and many hours of study about the various German aircraft equipment and systems, the time and date of the escape attempt was set. At the planned time and day, after evening roll-call, with the hallway kept clear of onlookers, the crew emerged from the barrack hallway, went over the fence on signal - four men in four separate jumps without anyone being injured by the fall and disappeared into the warehouse that was building #11.

Because I'd worked with the operation I was given a job as one of the two look-outs who would watch the distant tower guard to see when he looked in our direction. I sat with a newspaper I couldn't read as did the other look-out to send our signals. Newspaper open meant the guard was looking our way. Closed meant "go" before he looks this way again. All went well; the men moved like clock-work, faster than I would have believed possible. If I'd been British, I'd probably have said,"bloody acrobats they were."

We never saw those men again. We wouldn't know if they'd been successful or not. At this time of the war, if the escapees had been captured they would have been sent to a different camp unknown to us. This was to prevent our learning what had caused their capture - and to prevent us from building up our store of knowledge to use in future escape briefings to help later vagabonds.

Some years later, after the war ended, I encountered the captain, now a major, who had been the escape leader. When we recognized each other I asked him how the mission had gone. He told me that his group made it safely to the airfield at Barth and had waited through the preflight sequence I detailed earlier. Everything went as planned, until it crashed. When the single German crew chief completed his work at the plane and was picked up by a truck and taken away for breakfast, our crew waited until the vehicle was out of sight. They then climbed into the unlocked aircraft, a Junkers JU-88 twin-engine plane, and started the left engine which started immediately. The right engine, obstinately, would not *catch*, so they decided to taxi out of the hardstand with the power they had and attempt to re-start while rolling on the taxiway leading to the runway.

They tried over and over while taxiing, but the right engine, probably flooded, refused to power-up. The engine noises they made brought security vehicles out in pursuit to chase them down until they were surrounded by vehicles and guns. With a heavy load of fuel and bombs aboard they had enough power to taxi on the one engine, but not enough power to attempt a takeoff. The pilot stopped, shut down the left engine and with his "crew" climbed out with their hands held high.

The group was not mistreated or punished, but were probably secretly admired by the *Luftwaffe* officers and NCO's for having done so well and gotten so far. They were taken into the main operations building and briefly interrogated. It didn't take long, because none of the escapees would answer any questions. They were later fed before eventually being put on a train under guard for the trip to Stalag-Luft 3 at Sagan southeast of Berlin on the Oder River instead of being returned to our camp.

Later as the Russian Army approached from the East to the Oder River, our former *stalag*-mates also were forced out onto the road to march westward with the rest of the Pow's there. This was in mid-Winter in the freezing rain and snow in only the clothes on their backs and with only the few possesions and food items they could carry.

The German guards and the few officers left endured the same conditions even if they did have a few small wagons and carts to carry whatever small amounts of food and supplies that remained. After many days of misery, starvation, sickness, and many deaths they finally arrived at Moosberg southwest of Berlin.

It is a story few have heard, but it was every bit as bad as the Bataan Death March in the Philippines, but it was in Winter's cold instead of Summer's heat. As bad as the conditions were for the Pow's at *Stalag-Luft* 3, it was far worse for those Pow's who had the far longer march all the way from East Prussia at Heidekrug's *Stalag-Luft* 4. One group marched for 86 days without bathing, shaving, or even being able to brush their teeth. They slept in barns, fields, culverts, under bridges - anywhere, for shelter, with death marching only steps behind them. Many of the sorry details are covered very well in the book, *The Last Escape*, by two British writers. Read it if you can locate it. In the long run we were able to avoid that final misery.

In another interesting escape caper, a German *fatigue* or work uniform was made for use during a night attempt. To set the scene: each barracks had a white ball painted on the inside of a window covering shutter on a window facing a guard tower. These shutters were normally kept shut at night except for the very warm nights of Summer. If someone became sick during the night hours and needed to be taken to the medical clinic, this shutter was opened. As the tower guard's searchlight swept that area the light would reflect from the painted ball. In the Summer nights when the shutters were allowed to be open for ventilation, a *kriegie* lamp would be waved back and forth to get the tower guard's attention. The tower guard would then phone the guard station at the compound's entrance gate to report the problem and the barracks number. The duty sergeant at the guardroom located in the small building located between the inner and outer compound gates would send a one or two person security patrol to the proper barracks to investigate and determine what was needed. Each barracks already had a stretcher on hand for transporting patients to the clinic. If necessary the investigating minor German NCO would select the needed stretcher bearers from the men of that barracks to carry the patient to the medical clinic in West Compound. I don't believe the camp had any sort of ambulance - even horse drawn. I never saw one.

For this escape attempt, however, the shutter was not opened, the tower guard did not see the white 'ball,' and he did not phone the duty sergeant. Two POW's carried a third man on the stretcher and the group was accompanied by a *Kriegie* dressed as an imitation German security patrol guard *Unteroffizier*, or buck-sergeant German NCO.

The fake guard was a short and stocky German-American lieutenant whose excellent language skill was learned from his own family. His overseas cap was made from a similar RAF blue cap re-worked. Embroidered on the cap was the *Luftwaffe* eagle and swastika, plus the German national red, white and black roundel below it. The work jacket, also worn in hot weather, was sewn from white cloth, probably a sheet, with buttons carved from wood and painted silver with water color. No insignia, such as NCO silver *tress* braid around the collar, shoulder boards or collar tabs, was needed for the white work tunic, because none was normally worn on this jacket. The trousers were of RAF blue similar to the German Air Force blue color. The *jackboots* were made by sewing cardboard *uppers* painted black to RAF high-top black shoes. All this would probably not have passed inspection in daylight, but were fine for one-time night use.

The fake security patrol "guard" passed out through the inner gate, marched his group to the pass window at the guard house, tore off the coupon from the "home made" gate pass, handed the pass in to the duty NCO, signed the log, and moved his group to the outer gate. The escape group was allowed to pass through the outer gates and turned left down the main road past the camp headquarters toward West Compound and the clinic.

The escape attempt was observed by the tower guard who saw the procession which looked normal. He probably thought another tower guard had called in the security patrol man, but probably got *chewed-out* later in any case.

The escape attempt was timed for the period just after the eleven o'clock guard change so the incoming senior duty NCO, or his incoming staff, would not realize that no one had been sent to the barracks from which these men had come. The game finally fell apart when the newly arrived German duty NCO noticed that the gate pass coupon handed through the window had been torn from the gate pass. We learned later that this coupon was not torn off after nightfall when someone passed into or out of the compound gates. Evidently our observers studying the gate protocol from Barracks #1 nearby had failed to note the night procedure.

This pass was very simply made, and copied from a real one by me, with a blank line for the user's name and was countersigned by the German commander (also me). It had a perforated line made with a safety pin to provide a tear-off coupon with another blank line for the user's name. In this way the pass remained with the user and the coupon remained at the guard room, but only during daylight hours.

The duty NCO had been puzzled by the pass "tear-off" and realized he hadn't recognized the guard escort. To be safe he called the duty NCO at camp headquarters to check the name he had on the coupon. No man of that name was on duty or even known, so a small guard group was sent out to investigate. They intercepted the escape group still marching down the road.

Instead of spending the night in the nearest forest as hoped, the group spent the night in private rooms - in solitary at the klink. Their "trial" was held the next morning by the German *Oberleutnant*, the camp adjutant, who was also the military court officer. This officer had been a bombardier-navigator in light bombers on the East Front until his right hand was amputated by flak. We were told that he taught himself to write with his left hand during recuperation in the hospital.

The sentence for the escapees after their summary court trial was eight days in a solitary cell, or the klink. The fake uniform probably joined the German intelligence chief, Major von Mueller's, escape museum. Presumably it later was also brought back to the US by Lt Col Greening for his postwar Red Cross tour across America, which, unfortunately, I never caught up with or saw.

All sorts of escapes were tried. One of the Lt Colonels had himself wired to the transmission shaft of a supply truck. No, he didn't make it out of the camp, and unfortunately (again) he did get badly cut arms when his too thin arm and leg padding shifted when the truck lurched forward. He recovered in time and was somewhat wiser, if not happier.

A young captain unsuccessfully tried his luck by attempting to ride out of the camp in a large barrel full of unwashed tin cans headed for the German scrap drive. His barrel was loaded onto the truck by two very hearty young Americans. When apprehended he was dirty but undamaged.

The list of escape attempts when they were still permitted by our own command was almost endless. You can imagine how many ideas good and bad, smart, dumb, or batty can be thought up by very intelligent, generally well educated and otherwise, young men in the prime of health and physical condition, even if hungry. You can also imagine how many hare-brained schemes there were that were de-railed by the over-worked Escape Committee.

Late in the war a lieutenant, a loner, from our barracks made it over-the-fence-and-gone on his own during an air raid blackout. He had confided in no one, nor had he sought permission from the Escape Committee. He did it on the spur of the moment without planning or provisions. He continued on for two days, unstopped by anyone, then surrendered to police when he became too hungry to continue. He was brought back to the camp and probably welcomed the warm bunk and food in the klink

Another lieutenant, also late in the war, made it over the fence during the now numerous air raid blackouts. His plan, in his tea-dyed brown suit (from RAF blue, an awful color mixture that probably attracted attention), was to simply walk the main road toward the port. He'd claim to be an Italian road worker who didn't speak German if stopped by the police. We heard that the *Eytie* workers never bothered to carry their papers with them when stopped by police. First, of course, during the night he had to break into an unlocked house without a dog and steal a hat. It seems that all European males wore hats as had been true in America years ago when I was a boy. If you didn't wear a hat you'd probably be noticed immediately.

During the first morning while marching down the road while wearing the successfully stolen hat, he was stopped by the police. He pretended to be a road worker even though he didn't speak a word of Italian. He waved his arms wildly and shouted loudly,"Italiano, Italiano." It didn't last very long. There were actual conscripted Italian road workers not far away, who heard the commotion and told the police when asked that our man wasn't Italian. He was actually a well tanned American with black hair.

Again, because it was late war with little transportation available, he was returned to us after his guest appearance at the klink. Later he told us all the sordid details. We had a good laugh with him.

We could not escape by force even though we outnumbered the guards and prison camp staff, but we could try all sorts of mis-direction and stealth. Our best, and perhaps only chance, to escape was to find a very small, probably temporary, chink in the enemy's armor to wiggle through, before it was spotted and corrected by our keepers. Since the Germans were well experienced through many such wars they missed very little, stayed alert, and trained their personnel very well.

The German prisoner commands earlier during the war, as I've said, normally sent re-captured prisoners to a different camp in another area to prevent our Escape Committees from adding to their knowledge. Later these camp transfers had to be abandoned as conditions worsened in Germany and transportation became much more limited, drawn out and difficult, if not impossible.

As you'd imagine, when over the fence and out the gate escapes became well nigh impossible we began to plan tunnel escapes. Now when a tunnel digging plan was proposed it not only had to be sanctioned by the escape committee, but it had to be cleared by "civil engineering".

This group was lead by a pilot who'd graduated from West Point with a civil engineering degree, but who had transferred to the Army Air Corps at the war's approach to become a flyer. I don't remember his first name, but he was known by everyone, even the Germans, as Major 'Mole' Palfrey. Our small peninsula, projecting from the mainland into the waters of the Bodensee on three sides, was only seven and a half feet above sea level. To begin a tunnel we dug straight down until we hit the water table, then horizontally outward in whatever was the shortest distance to an area outside the main fence that provided a safe spot to dig upwards toward the surface.

Originally we could tunnel from under various barracks to the East, North and West. South was out, since that was toward our mess hall and the German barracks area. Later, when North Two compound was constructed, that cut off the North side. That left East and West and depended on what side of the compound your barracks was located. The East side was to a potato field - but it was too obvious and had been used earlier, with the current Senior Allied Officer in the escape group, soon after our arrival at the *stalag*.

Since I now lived in barracks #10 on the West side, our shortest route was to and under the heavily traveled concrete two-lane road previously mentioned. Over this road traveled very large trucks, often towing the heavy Flak 36/37, 88 millimeter, anti-aircraft cannons or the Wurzburg flak radars and their large power generators. The German Flak school south of our camp held regular field practice for students in the wide meadows north of the camp to teach the crews how to use the guns in the open, or to use their cannons as anti-tank weapons in the way that Rommel had pioneered in North Afrika. These large units trundled by on a regular basis, since this was the only available paved vehicle route from the Flak school at the southern end of the peninsula to the training grounds for the Luftwaffe crews at the far north end of the peninsula.

The sandy shallow soil beneath us had a high risk of cave-in danger, but was easy to dig through. Our *shovel* was the ever-present Klim powdered milk can from the Red Cross food parcel. It was small enough to be used in the very limited tunnel space and it worked well with the soft, damp sand. For shoring we had to cannibalize the barracks buildings for the needed wood bracing. If the tunnel had been much longer, we would have had to take so much wood from the building that it would have been in danger of collapse. The mouth of the tunnel shaft was usually the rectangular wooden frame that held up a table top - from one of the two in each large room. When digging stopped at night, boards on a table top closed the opening and was covered with sand to prevent discovery by the *ferrets*. These were small German guards from the security patrol who searched the crawl spaces at night under the barracks for signs of tunnel digging activity.

Closing down our tunnel dig in this way allowed us to easily and safely shut down digging for the day and then re-open the shaft more quickly the following day.

We needed to take great care in the planning of and the actual construction of tunnels under these soil conditions, because we couldn't have rescued the victim of a cave-in quickly enough to prevent death from asphyxiation - or even from being crushed by the great weight of damp sand. Of course, if the sand was too dry its lack of cohesion would have made digging impossible at any depth.

The digger at the tunnel face piled sand into a wooden coal box that sat upon wooden rails behind him that were added to as the tunnel lengthened. When the coal box was filled it was pulled back to the vertical shaft, detached from the cord and lifted up to a third man above for disposal. The lifter then sent the empty box forward pushing with another wooden strip. The process worked efficiently, if slowly, but was certainly no job for the lazy - or for someone bothered with claustrophobia. Fatigue, however, was a major factor. Diggers had to be regularly replaced in an hour or less. Even being small and moving only a few pounds at a time, I was soon covered in sweat on each shift, regardless of the season. Perhaps an unknown factor then might have been apprehension as we slowly dug or backed toward the vertical shaft - there wasn't room enough even to turn around.

Another problem that had to be recognized and countered was the low tunnel air supply. Even at this shallow tunnel depth we found that the digger, due to his exertions, needed a greater amount of air than the tunnel could supply. A few men passed out before the lack of air was understood. To provide a greater air supply a table top was used with a cardboard bellows added inside of the "box" under the top. After being hinged, the table top was lifted and pushed down to act as a pump. A table leg was hollowed by burning out its interior with a hot wire from the room's tile stove - our usual drill. It was slow work that took a lot of time, but helped fill our many long days. We had a large supply of free and enthusiastic manpower, plus unlimited time, the two most needed ingredients. To carry the air to the tunnel we had a removable floorboard plus lengths of biscuit tins that had been fitted together by the tinsmiths - usually in three or four foot lengths that were added to as the tunnel lengthened. By cutting off one of the rims the cans could be squeezed tightly together and held in place by friction. Some tunnels used a floorboard with a hole in it for the air line that had to be switched before room inspections by the German security patrols. You also had to color or "wear" match the replaceable floorboard so it wouldn't attract any unneeded attention from visitors.

Each tunnel enterprise required far more people, labor, materiel, and time than we at first expected, or planned for. You can also see why we needed the goon guards at either end of the barracks hallway to alert us to approaching German security patrols or even non-Allied visitors.

A tunnel plan took many days to plan and to get the multi-level approval needed from our escape committee and senior officers, plus the promised assistance from all the "specialists" who would make the air lines and other items to be used. It took even more unnumbered hours by many hard-working men to complete. Fortunately we had all the volunteers we could use. Disposal of sand led to many solutions - some hilarious. One of the most common methods used long thin bags inside trouser legs and attached to a belt. The sand was carried to many nearby construction sites, such as the building of volleyball courts or other large-scale jobs such as the re-building of the softball diamond. When the hauler arrived at the dump site he pulled a string at the waist to open a slip knot at the ankle. Then he had to move around so he wouldn't leave a noticeable pile of sand high enough for a tower guard to notice and associate it with him. Even with a group of men hauling sand it was a slow process that did not move much sand quickly enough. Now I realize that the German probably had a good laugh at each of our many attempts. Like kids we burned a lot of energy, but we didn't get out - all we got was dirty. Farther along I'll tell you why.

My only tunnel work was from a tunnel that began under my lower bunk against the room's outer wall. It was through the floor to the crawl space beneath and to the table top shaft entrance one foot under the sand surface directly below. After the irregular-shaped floor piece was lifted the sand-filled coal boxes were passed up through the open space where my mattress had been rolled to the bunk's lower end. The bed slats that remained after my model-building efforts were pushed back to open the space. The man who received the coal box in the room took it out the door to the corridor and up the hall to the next room where there was a ceiling hatch. The hauler stepped up onto a bench and handed the box up to a man in the attic hatch under the low pitched roof. The attic man, with a *kriegie* lamp and a book to read, took the filled box, emptied it, and lowered it to the man who'd given it to him.

This went on well for a time, but suddenly there was a large crash. The whole ceiling section in the room with the ceiling hatch fell with a cascade of sand that covered almost everything in that end of the room. The dummy in the attic had been piling the sand in one large mound - not spreading the dirt out evenly as he'd been told to do. The ceiling couldn't support the weight of the load of sand all in one place. This caused the initial collapse and the immediate problem of disposing of the sand spread across several bunks and a quarter of the room.

Like a college prank this caused a mad scramble. The tunnel-digging was shut down, the shaft was closed and covered - and a frantic operation was begun to nail the fallen ceiling in place again, to gather the fallen sand and spread it evenly and properly in the attic, then clean the room to make it look normal before a German security patrol blundered by.

The word was passed quietly and a steady stream of volunteers arrived individually. A sudden rush with a gang carrying buckets and brooms would have tipped the tower guard nearby that something unusual was happening. This "accident" happened in the early afternoon, but through fast and effective efforts by all who helped, I can testify that the room looked almost normal before I left in time for evening roll call. I wonder now, if a passing German would have thought that the room looked too clean and orderly.

One funny note came up later that evening. An upper bunk had filled with fallen sand at the ceiling collapse and then in the rush the bunk had then been hastily covered by a blanket. Evidently no one remembered and returned to clear out the sand from the bunk. In any case, the bunk's occupant returned from his day volunteer job across the compound at the library. He'd been told about the afternoon's frolic, but again no one remembered the sand in the bunk. After "lights out" at ten, this individual climbed up to his bunk in the dark, pulled the cover back, and slid into the mass of cold, wet sand. He screamed and jumped out into space. He continued screaming until someone lit his *kriegie* lamp to find out what had happened. Now, finally, it was realized what had been overlooked. After the laughing died down, several roommates volunteered to clean out his bunk the next morning after roll call. He said it felt as if some animal had grabbed him in the dark. He had to spend the night sleeping on the floor. The story was repeated over and over in the other barracks and probably in the other compounds long afterwards. I still laugh internally whenever I remember.

The next day a German security patrol came to our room accompanied by an officer. He pointed at my lower bunk and asked whose it was. After stalling we finally admitted that it was mine. I was marched across the compound and out the double gates down the road to the German headquarters. There the camp adjutant, who was also the summary court officer, told me in English, that since my bunk was involved in the attempted tunnel escape, I would be charged with that offense. He asked if I had anything to say. I told him I didn't. He sentenced me to eight days in solitary. He didn't also charge me with *destroying the property of the Reich.* which would have been an added eight days and could easily have been added, because all the materials used in the tunnel building had been torn from the barracks or had "modified" the furniture.

Normally you served eight days, got out for two days, then went back to solitary for eight more days if that was part of your sentence. This system was also used for longer sentences, such as Col Russ Spicer's six month sentence. He'd originally been found guilty by an international court of attempting to incite a riot while under confinement - which is a capital crime under military law in Germany. Instead of a firing squad he was shuttled in and out of solitary for about six months.

I was walked back to the compound under guard by two German NCO's and was allowed to collect my overcoat, a blanket, a book, and my pocket knife for whittling, plus mug and spoon (plus the P-51 model I was making). Then it was back to the klink.

The solitary block was an "L-shaped" addition to the guardroom at our compound's entrance and built after our arrival. Prior to that coming on line, we had to share the one in West compound. I was taken through the duty sergeant's office and then down a corridor to a cell. The small room contained a chair, small table, and a single bunk similar to my usual *lower*. There was a single overhead light with a wall switch, but no real window. That opening outside of the glass was covered by an angled board that would allow daylight from above, but wouldn't allow you to see outside. The heavy wooden door had a small peephole with a metal cover on the outside. Here it was called a *Judas Window*. The guards could look in, but you couldn't normally look out.

Just before dark a guard opened the cell door and showed me to a small side door which led to an enclosed *dog-track,* where I could walk for an hour. Well after dark the same guard came back to return me to the cell. Only one person at a time could be led out to the dog-track. otherwise it wouldn't have been *solitary*. Soon after return the guard again opened the door and I was handed a tray with potatoes, cabbage, black bread, and ersatz black coffee with sugar.

After eating, I read for some time, then turned out the light and fell asleep. The eight days went by quickly. I loved the peace and quiet after living in a 12 x 15 foot room with eleven other men who talked all day - and snored all night. I walked my hour outside each day, ate the same German food I'd have gotten at the mess, and had all of the uninterrupted time I wanted to read or whittle on the model I'd brought. I didn't have the *extras* I would have had from the last Red Cross food parcel I'd been given at the barracks, but I knew that what was left there would be safe with my roommates until I returned. The honesty I saw among the prisoners I knew was outstanding. It was unlike anywhere I'd seen before or afterwards. Along with the helpfulness and cooperation, life was very different there then usual.

What I soon learned, also, and gradually spread around the camp, was that solitary, or the klink was a mini-vacation - that instead of a punishment it was actually a reward. As time went by, and only two solitary blocks were built, first in West compound and then later in North One compound, the number of men sentenced to eight days or more gradually grew. The waiting list lengthened until you had to wait weeks, then months, to get a *private room*. We even joked that you could get a private room with *limited booking*, as the RAF chaps used to say, if you could manage to get into the always crowded solitary block. My only other vacation came with eight days in the Fall of 1944. I threw a rock through a window while the *kommandant* was walking past. *Destroying the property of the Reich* was the charge, but wasn't that the mission of the 8th Air Force?

The biggest attempted tunnel breakouts from our North 1 compound was in the Spring of 1944 when we lived in Barracks #7, in the middle of the compound, soon after we arrived. It was the one led by then Senior Allied Officer, Col Jean R Byerly, who went with a group of 60 men who tried to leave by tunnel from under Barracks #4. Their exit was in a potato field outside of the eastern fence. All through that selected day those leaving gradually collected inside the barracks with their traveling needs so they wouldn't alert the nearby tower guard.

The next vital job was to prevent the holding of the evening roll call, or *appell*, so that the absence of this large group wouldn't be noticed until they dispersed and got out of the camp area. The riot was begun by Flight Officer (WWII warrant officer) Bob Hoover, jack of all trades and the famous aerobatic pilot of later years and "chase" pilot for Brigadier "Chuck" Yeager.

After we formed up in our usual parade formation for the roll call by barracks, and just as we were called to attention, Bob threw a large, wet, and steaming piece of horse manure smack into the back of the head of our barracks leader (the same captain who later led the "aircraft stealing" attempt at Barth Fliegerhorst, and was trained by Bob Hoover). This "missile" had been delivered just before roll call by the horse pulling the wagon that delivered the coal briquets to the compound. The tall captain wheeled around and floored Bob with a single roundhouse punch. I don't think he was in on the gag.

On cue, spotted around the ten squadron formations, were others whose job it was to fiercely smack an unsuspecting man next to him. This created many real fights when those not in the know angrily fought back. When one area simmered down another would start up anew. It was quite a scene. Only a very small group knew what was happening, and they didn't know why, but they pitched in as instructed, or should I say, *punched in*. People began fighting everywhere. It went on and on, regardless of what the German shouted at us.

Finally the German major in charge no longer tried to conduct the roll call. He gave up and called his men to clear out, except for those older men who guarded the barracks end doors across the compound. He ordered that the barracks be locked down for the night. We couldn't get back in to our rooms and would have to sleep on the ground in whatever clothes we were wearing. I was fortunate enough to show up for that evening's roll call wearing my brown Belgian officer's overcoat. I also stood well clear of the battlers. No one punched me. We didn't get dinner either.

Unbeknownst to us at an hour or so after nightfall the sixty escapees poked up through the potato field where their tunnel ended. As the first man, Col Byerly, emerged, the northeast tower guard turned on his spotlight and illuminated the exact spot, while a large group of guards with rifles and machine pistols quietly surrounded the tunnel hole. All of this was some distance out into the field from the *stalag* fence. In the compound another group from the security patrol surrounded the doors of barracks #4 to arrest all those equipped for a going-away party. Evidently the Germans were a lot sharper than we expected. They seemed to know all the details. Most of these details come from hearsay, since I slept straight through the night part and woke at sun up. Later I saw many split lips and black eyes, but none of the many body bruises generated by the ruckus. Many awoke unsure of what had really happened. I'm not sure how the Germans were able to handle that many men in the klink, but gradually we got most of the details of the fiasco when the would-be escapees filtered back to the compound. The forgotten guards, the old men who guarded the barracks doors, also had to maintain their posts until relieved. If they were changed, it too was while I slept through the night.

We later learned that the staged riot was well known to the Germans if not to us, and was used so often in so many camps that they probably had to stifle a yawn when they saw it repeated by us. Instead of hiding something from our keepers by fighting we were probably alerting them.

Here is the secret that I alluded to earlier when I said, "farther on I'll tell you why." It wasn't until the end of the war, and after he found out, that Col Greening told me that he'd recently learned that the Germans had small seismographic devices planted around the camp's compounds to locate our digging and find our tunnels. I never saw the plate-sized sensors, but I'd still like to see one.

Another potential escape attempt that never came off involved the coal briquet pile where the fuel was sorted before being transported to the compounds. A small group of men from our barracks room, led by B-17 pilot, Lt Mike O'Shea, volunteered to sort coal briquets at sheds near the woods area north of North 2 compound. North 3 hadn't been built yet. It was a way to get out of the compound and get exercise while looking for escape possibilities where there probably wasn't much guard over-sight. The idea grew that men could drift away and not be missed for some time. They'd probably seen too many old-time chain gang movies. Since I wasn't involved personally I don't know if this plan was sanctioned by the escape committee. If they had maps I wasn't aware of it, since they did not come from our map and document operation. I do know that the men filled their overcoat pockets with as much non-perishable food as possible each time they left for work. Day after day drifted past, the men came and went to the coat site, with no opportunity to escape. All they ended up with was exercise, muscles, and lots of coal dust.

The German intelligence officer for the camp was Major von Mueller of the Luftwaffe. He told me that before the war he'd lived in Santa Barbara where he'd raised polo ponies. I thought that this was a lie at the time, but I later learned from a Santa Barbara acquaintance that his statements were true. I also heard that he'd been shot and killed by the Russians soon after they arrived. Finally I learned from the camp's historian, Helga Radau, during one of our several visits, that von Mueller lived and later died of old age under the long Russian occupation in Barth.

Throughout their history and the many wars they've engaged in, the Germans have made a wide study of prison camp security. Camp authorities certainly knew more about keeping prisoners in than we knew about getting out. A few men got out, though for one reason or another, but no one ever got away.

Col Greening had most of the same escape models and related material packed and returned to the US when we departed for study by our intelligence people, He also had a model of our North One compound made up by using cardboard from the large Red Cross cartons. I saw it during construction. It was very complete and accurately done with each of the barracks and their locations. The barbed wire was made from black thread with many fine knots to simulate the barbs. I hope it is in some museum where I might discover it.

The main reason for having an escape committee with firm control of such attempts is to prevent conflicts that would interfere with an operation, or operations already underway, or ones that have a better chance of success. Well planned and coordinated efforts have far more chance of success, though escapes in general have a very low success rate. Training, briefing, and a supply of the needed items are essential. The loner who climbed the fence during an air raid blackout and somehow managed to wander away carried no supplies and hadn't even learned how to find the necessary food in the fields to keep him going. He did have *one man security*, since he was the only one who knew he was going, but this is only one of many essentials for success.

We've all seen many fictional movies and have read related stories about escapes by war prisoners, but virtually all are merely fictional - not reality. Escapes are not impossible, but are highly unlikely when the guards and leadership personnel are well motivated, highly trained, and well led by intelligent NCO's and officers.

The Berlin and the local German command steadily posted many dire warnings and threats on our bulletin boards about what would happen to re-captured escapees when caught and about the futility of our attempts. Their claims grew more strident as the war progressed, but our efforts and activities, even if ineffective, did not wane. This is probably true of most 'caged creatures.' especially young American aviators.

No, I still haven't thought of a successful way to escape from the Stalag Luft camp, but in the long run we were probably safer there than out in the mad house Germany became.

13
Invasion

The Invasion of *Festung Europa* (Fortress Europe) - The German government knowing very well that the Allied invasion was imminent, announced in the press and over the radio many times during the Spring of 1944, that when the great attack began all the air raid sirens would blow and the church bells would ring for ten continuous minutes to warn the people and to call everyone "to arms". They never mentioned that even to the German people if would really signify a great relief. The only invasion fear they had was from Stalin and his Russian hordes that had planned to overrun Europe as early as 1941, before Hitler unknowingly beat them to the punch with Operation Barbarosa.

In *Stalin's Folly*, a book by Russian professor, Constantine Pleshakov, tells clearly of the 149 Russian divisions forming on the Polish border in preparation for the attack against not only Germany, but to conquer all of Europe. The pre-emptive strike by German forces during Barbarossa captured and killed more than three million men and resulted in the capture of vast amounts of military materiel. No, the American press or even the American government has never bothered to tell you about this, believing or pretending that Russia was and is an ally. Having flown the majority of my military years with our Strategic Air Command, I can tell you a vastly different story.

Playing upon this event, which we all hopefully looked forward to, since to us it would bring the war's end closer, many of the German guard NCO's whispered in the mornings and evenings from quiet corners to American and RAF officers that the invasion had started the previous night.

At first we gladly believed them, since we'd seen the earlier newspaper warnings and articles about the imminent battle. We'd be elated, believing that we'd be leaving our German "home away from home", soon. Then would come the gradual deflation as our hopes leaked away like a balloon with a tiny hole.

Yes, we got day-old copies of the local *Stralsunder Zeitung* and other papers that were brought in by helpful NCO's and officers on a fairly regular basis and we eagerly listened to everything read out to us. At first we believed them and helped them spread the rumors, because we believed that the war's end couldn't be far away due to the Russian advances and the German defeats on many battlefields.

At first we were gullible and we'd believe (or hope), then become disillusioned by the lack of follow-up information. Finally after several false starts we gave up and wouldn't believe anyone's story or rumor. Then no one would listen no matter what the source. The bubble had burst.

On the morning, of June 6th , however, I awoke soon after sun up, before five A.M, dressed, and went out for a walk around the compound. There was a beautiful blue sky and a mild sunny morning. My route around the fence was 1/3 of a mile, by our estimate. Being so far north, and on the Baltic Sea coast north of Berlin, the sun barely slides below the horizon during the very short nights at this time of the year. I can see why the Teutonic people dearly loved their mid-Summer celebrations after enduring the long cold and dark winters. The dreariest days back then made the Summer sun's blessing more beautiful by comparison. Also we lived in a beautiful and almost unspoiled natural setting - with the sparkling, shining seas around us counter-pointed by the tall dark pines. The only real blots on the picture were our barracks buildings. Now, we have seen during our several later visits, after many years of war and forty years of foreign occupation, the area is back to its earlier more beautiful times.

As I walked out through the un-gated opening that separated our barracks row from the main compound, I met the German compound commander, Major Schroeder, and his senior NCO who were coming in and headed for the barracks where our senior officers were quartered.

The Major asked, "What are you doing up so early?" I answered that it was too bright to sleep. He grinned and said, "The invasion started last night in France, with your *fallschirmjaegers* (paratroops) landing in towns along the Channel." I said, "Not you, too." He told me that it was really true this time, and that he'd bring the newspaper when he came back at nine. Neither of the men seemed to be unhappy about the coming invasion, I'm sure they wanted to see the war ended. Then the two men went past me and on to Barracks #9, the senior staff barracks to pass the word officially to our Senior Allied Officer (SAO), Col Hubert L Zemke, his staff members, and the Group One - North One compound commander, Lt Col Ross Greening..

Instead of continuing the walk around the compound, I hurried back to our room #7 in barracks #10, woke all my roommates, and told them what I learned from the Major. No one would believe him or me, or even listen, after all the previous disappointments. Then I began to bet with the non-believers everything I had in the way of hoarded food. The bet of a piece of cheese, bread, ersatz coffee, sugar, whatever (our local currency), was quickly accepted. When I saw that I wasn't getting through I stuffed the notes in my pocket. I gave up, and made several circuits around the fence line. Even at roll call and breakfast no one seemed to accept that the invasion had actually started, even when I quoted the German compound commander.

Near nine, though, Major Schroeder, as good as his word, brought the local paper. There was a huge headline, about two inches high, on the front of the *Stralsunder Zeitung* newspaper which the German officer brought. The hastily written articles, probably taken directly from the German radio reports were about where the American and British forces were landing. They were quickly interpreted by our many German-speaking prisoners. Berlin radio, of course boasted that any of our forces that survived the landing would quickly be pushed back into the sea.

I'm sure our senior staff officers and Senior Allied Officer, Colonel Zemke, playing their cards close to the vest, had picked up the momentous news long beforehand from their BBC radio reports being secretly received at all hours, and had waited until the Germans acknowledged the event without revealing what they had received prior knowledge of. With the newspaper verification and the realization that the end was finally in sight, the uproar began. I can look back now and imagine how great an announcement like this was received by the RAF long-termers in West Compound.

Now I was suddenly rich in Pow currency, food, as I took a big Red Cross parcel box around to claim my winnings. Even the losers were happy to pay up. No one reneged. For the next few days we were all on cloud nine. The gabbing never stopped. We tried to happily out-talk each other non-stop.

One of our first occupations after the invasion started was to begin a large, hand-drawn wall map on the only wall in the room not covered by the two-decker bunks, the tile stove, or the tall closet. We even had to move the large closet to get a wall surface big enough for our needed wall map.

Lt Arthur "Herky" W Starratt, a B-17 bombardier from Maine, went over to the library where he volunteered, and brought back a book which had a small map of Europe. It ran from the Atlantic to well into Russia. He asked if I would scale it to fit the large wall space so we'd have enough detail to plot the Allied advances. The little map grew from 6 inches high to six feet high over the next day or so. There were more helping hands than we could use, but it was all worth the effort.

With the drawing tools I'd made when I'd arrived at the camp and later used in my drafting classes, I now used them to scale our large map with long-hand arithmetic. The ruler was a great asset in laying out the little grid on the smaller map and then translating the map details to a larger grid on the wall. The dividers, made originally from two safety pins, were also a great help and speeded the work. We not only laid out the western Europe area, but were able to stretch the map to cover the area where the Russian front was located in eastern Poland. We also drew our own area more accurately from details derived from the secret escape maps no longer needed and could finally see in more detail where we were. In places where we had little detail we used what we called "artistic license", or "fill-in-later".

Other maps were found in library books. Several were even in the newspapers to show not only France but battle areas in Poland. The wall board was like the our US type and was easy to draw upon. When the large scale map was finished, and was up-dated with each new scrap of information from both the east and west fronts, we had a continuous stream of visitors to see how the war was going. We felt like the local news bureau.

This map gave us a continuing view of what was happening on each front, and was a real help during the later setbacks during the "Bridge Too Far" and the "Battle of the Bulge".

Because of our Allied intelligence errors and failure to properly do reconnaissance patrols behind enemy lines, that last-gasp counterattack foundered only on the enemy's lack of fuel for their armored units. In fact, though, it would have run out of steam in the long run as the German forces ran out of every type of needed supplies, fighting men, and national morale.

Our room became the most popular in the compound. It also encouraged many other barracks rooms to begin their own large map to keep up with what now became the daily news. The German press and radio gave accurate accounts, although the Allied press had always denied that. The map continued to get more detailed until we finally walked out the gate in May.

Each of us was dreaming up and discussing with anyone who'd listen what we'd do when we got home, what we'd eat, and everything remotely imaginably. A few even began designing houses they'd like to live in. Just like kids before Christmas. Even as skinny as we were, you'd have gained a few pounds even listening to those vocally planning the big high-calorie meals they'd have upon returning to the US, or at stops along the way.

As things worked out, though, we had another very long eleven months to sweat out before the Russian Army showed up in the form of the Russian 44th Guards Rifle Division, commanded by a General Borisov. During that long time, almost a year, North Three compound was erected and a great many more prisoners arrived from camps over-run in the East. Even the Russians stalled us for ten days after they arrived and after the war ended before we were allowed to leave. It was not the end, only the beginning of the end. The war had still to be won.

14
LIBERATION

We had plenty of advance warning of our impending liberation, although we had no idea how events would work out. The earliest sign of the approaching battle front was the increase in flight operations from our nearby Barth airfield with droves of *Luftwaffe* fighter-bomber aircraft such as the Junkers JU-88 Schnellbombers and the Focke-Wulf FW-190 Butcher Birds. Other planes, such as Heinkel 111's, Dornier 17's and 217's and anything else that could be pressed into combat service joined the battle in the east. They each took-off on ground attack missions trying futilely to stop the oncoming horde. We heard them around the clock, leaving and returning.

At the same time, other workers, unknown to us, at the airfield were hurriedly trying to assemble the Heinkel HE-162 Volksjaeger (People's Fighter), a small twin-tailed, single jet-fighter. Parts for these planes were being made at five SS-labor camps to the west and south of us. These were the sub-camps of the infamous Ravensbrucke main concentrations camp farther south of us mentioned later in the Chivalry chapter. We learned nothing about these camps until after the liberation although we had seen gaunt prisoners in striped suits marching along the road on the outside of our compound. These men were probably from the SS camp at Barth airfield who were sent out to labor on our peninsula. We knew by the look of them that they were not out merely for a stroll, even though we didn't know who they were. They looked like very long term convicts.

The Barth airfield was close enough, on the south side of the camp, so that we could see the aircraft tails when they taxied by. We used to count out the numbers leaving and later returning to see how many were lost. We even argued with the tower guards who always claimed a higher return number. They insisted that their increased height gave them a far better view and more accuracy, but even they, as all of us, I'm sure, could read the writing on the wall. Good-natured heckling was the only weapon we had left. It finally felt very good to be on the winning team with the game winding down. No overtime!

The next and more urgent phase began when our ears picked up the distant and very faint sounds of artillery far away when the wind blew from the East. As the battle front crept ever nearer day by day we could gradually hear the artillery louder and nearer - and hear the guns regardless of the wind's direction. The roar gradually grew in volume and continuity, almost like the surf rising with a large storm approaching, which it was. At sunrise and sunset, and all through the night the ever-increasing thunder brought fear, anxiety, and probably a certain fatalism to the local military and to the nearby townspeople as they awaited their future, wondering what fate would bring them. I'm sure that anyone who could leave for the West did so. To all of the German people waiting nearby the fear of the oncoming Red Army horde had to be very much like the Huns awaiting the onslaught of Genghis Khan and the other barbarians with their droves of savage steppes people who had slaughtered so many of their ancestors in ancient times. At the same time it sounded to us like the sliding of the bolt that was unlocking the gates of freedom.

Toward the last days we began another large drawing on our room's walls. This added to the large wall map mentioned earlier. This was a large list we called the "Shitzen Liste" to record the names of the camp un-friendlies. The German NCO's believed it was a list of all those who would be shot after we were liberated. Actually it was for propaganda only. We don't really shoot people, but it did cause our guards to reconsider their past actions and possible futures. One of the *feldwebels* (Tech sergeant equivalent) heard about it and did come to our room to see if his name had been added. He now smiled rather than frowned. No one gave him any assurances or explained.

As the last days neared, the German *Kommandant* had delivered to us very large "drain tiles" of about 8 feet in diameter and 6 feet in length so that we could dig emergency water wells and be assured of a water supply. To dig you just put the pipe on the ground , open end down, wherever you wanted a well. Then you climbed in and shoveled the dirt over the top. The pipe goes down with you in it, easily and safely.

There was no shortage of volunteer labor. Many of us wanted to take part. I guess at our young age we wanted to be part of any new venture as ours spirits rebounded, then soared. The job was a quick and easy method of well drilling. The water table on our peninsula was only 7 ½ feet deep, since we were surrounded on three sides by the waters of the Boddensee, so it didn't take very long for the six foot section of drain tile to disappear into the ground. We later had to filter the sand from the water through cloth and then boil it, but it was never actually needed. We were told by the German NCO's that the Russians would probably cut off our electrical power when they overran the nearby town of Stralsund to the east where the local power plant was located. In retrospect, I don't believe they ever did. They saved it for their own later uses during their long 40-year occupation.

In the last days we were also allowed to dig slit trenches along the side of each barracks to protect ourselves from possible Russian artillery shelling as their troops approached. The Germans even provided shovels and allowed us to tear down the warning wires next to the barbed-wire fences to make more room. After all, we were a military camp and might not be recognized from a distance or from the air as a prisoner of war camp, even with the guard towers, soon enough during the Russian army approach.

We knew the war was all over but the shouting when the German staff began outdoor burning of field manuals and piles of their records and classified papers.

During the last days before the German departure the demolition crews from the flak school began using electrical fuses to detonate small aircraft bombs individually under the rows of Wurzburg flak radars in their storage garages to keep them from falling into enemy hands. We could hear the loud blasts and see the clouds of smoke and flying debris. There also were isolated small ammunition storage sheds sunken out in the nearby fields, safely apart from the guard barracks, that were destroyed in a similar manner. In one blast nearby, we saw the unexploded stick grenade explosive cans and wooden handles flying separately through the air. Cases of the stick grenades were stored with the explosive cans not containing explosive detonators installed and not screwed to the long throwing handles. There were also many red-painted wooden training bullets flying through the air. The grenades weren't dangerous, because they could not explode without their detonators installed. From our North 1 compound we were nearer to the action than North 2 and 3 and could hear more of the loud explosions and see the smoke plumes that marked their progress.

The large, wheel mounted, radar units, now being demolished, had been used to train air defense crews at the nearby *Luftwaffe* Flak School, just south of us. The original flak school at Munich, we were told, had been destroyed by a British night bombing attack. These Wurzburg flak radars' details, the forerunners of our American army flak radars, had first been captured for study purposes, as mentioned earlier, in several cross-Channel raids by British commando led attacks at the successful Bruneval and the Dieppe raids on the French coast. However, the only part of the Dieppe landings that was successful was the capture of the electronics from the longer-range Gemma coast-watcher radar. The Wurzburg radars had been used very effectively throughout the war when teamed with batteries of the famous German 88mm flak guns.

A woman we later knew in St Petersburg, Florida, had been a *Luftwaffe* phone operator with flak units back then. She called data to the gun crews during battle.

Quoting Lt Donald M Menard, in his report of the air evacuation of Stalag Luft 1, "Due to the proximity of the Russian forces to us in late April, the German staff asked for a meeting with Col Zemke. The purpose of the meeting was to propose turning over command of Stalag Luft 1 to our leaders while the German personnel, in civilian clothes, headed westward to contact British airborne forces who were within 70-80 km of Barth. The German colonel didn't want his men dying needlessly in a lost cause when the war's end was so near. This was agreed upon. At approximately midnight on May 1st, the German departure took place. The camp population was not aware of this operation until daylight when they awoke and saw that the Stars & Stripes and the Union Jack had replaced the swastika on the camp flag poles. The flags had been sewn with scraps and hidden for quite some time."

Col Hubert Zemke gives a complete description of his refusal of the evacuation order from Berlin to the then Camp *Kommandant, Luftwaffe Oberst* von Warnstedt in his book, *Zemke's Stalag*

Another version of the *stalag's* surrender to the Allied prisoners' command started with a much more somber and threatening beginning. This explanation was learned only after the war from Col Greening when we met again at MacDill AFB before his untimely death in 1955. His version is much more likely and believable under the circumstances. It began with an order from Berlin to the German *Kommandant, Oberst* Warnstedt, to evacuate the camp and march the prisoners westward under guard.

The *stalags* in eastern Germany and Poland had already being evacuated ahead of the Russian approach and marched back into Germany at a huge cost in death, disease, starvation and every privation possible. It is a largely unknown story in America and Great Britain, but was every bit as bad as the Bataan Death March that has been so well publicized. No comparable numbers of dead American and RAF flyers has ever been published to my knowledge. I have met some of these survivors and know the truth. This story is also covered very well in the book, *The Last Escape*, by a pair of British authors. I have a copy and have read it.

The Berlin order to *Oberst* Warnstedt was in connection with a Hitler order that directed prison camp commands to march prisoners to the west when the Russians approached. His order further added that if prisoners could not be moved in time they were to be shot, not released.

According to Ross Greening, Col Zemke refused the German order to prepare for evacuation. Instead he is said to have told the kommandant that with only the two hundred or so guards, and the almost ten thousand of us, the guards might shoot many of us, but then we would overwhelm his soldiers and few officers and kill every one of them.

This version is also spelled out very clearly in Col Zemke's postwar book, *Zemke's Stalag*. Apparently the German command chose the wiser, more survivable course, and turned the camp and small arms over to us, then left in peace in civilian clothes. I've never heard any more about them. I supposed they were released from their group and went their own ways.

The Senior Allied Officer, Col Zemke immediately put the previously long-planned "Field Force" into action. One of the first acts was to send out scouts to the east who spoke Polish or Russian. There they dug-in and awaited the advancing Russian infantry reconnaissance to prevent an inadvertent attack. Col Zemke had been a former American air-attache in Moscow and spoke Polish and Russian to some extent.

The German commander had stored a large pile of Red Cross parcels for our later use, so many at a time, in his classified storage garages behind the Wurzburg flak radar units at the Flak School. Now they were released to us, undamaged by the demolitions, so we'd have an adequate food supply until we were evacuated to the west.

Col Ross Greening, our compound commander, asked if I'd like to help. He gave me a group of 200 men to march to the food storage area, to retrieve the 44 pound boxes containing four eleven pound parcels each, then carry them back to the compound for safe keeping. We continued our shuttle until all the food was safely stacked in our barrack's central hallway. There were several other such groups from the compound sharing the load. Col Zemke also noted in his book that after the German Luftwaffe forces left, civilians unsuccessfully came to take this food for themselves. They backed off when faced by our now armed American forces.

To maintain order and safety for the prisoners and local population, our armed Field Force groups were sent to take control of each nearby area and road intersection, including the town, the airfield, etc. The departing Germans had provided the needed small arms such as pistols, machine pistols, rifles, ammunition and grenades.

After the ration shuttle I became a guard at an intersection not far from the camp in a largely agricultural area. I was armed with a GEW-43 machine pistol, ammunition clips, and a grenade bag. For identification I had an armband with an American flag, prisoner dog tags, plus my American uniform and insignia, such as it was. Previously I'd embroidered an American flag on the left shoulder of, first, my RAF uniform gotten at the transit camp at Oberusal, then, later, on the American uniform I'd traded for. I also wore the tea-stained second lieutenant bars. Today that get-up would look like a very poor re-enactors outfit -"not ready for prime time".

Not long after I took my post I was approached by an old German man who came to shake my hand, though he spoke no English. He was followed by a woman from the nearest farmhouse who asked me, in bits and pieces of English, to move into her family home when I was off duty to protect them from the Russians soon to arrive. The older gentleman was a nearby neighbor. At the home there was a grandmother and two wives with three children. Both of the soldier husbands were missing in action on the Eastern Front.

When I saw Ross again I asked permission to do so. He told me to do what I could for them. I spent several nights sleeping on a sofa at the farm. With his permission I also brought food for them from camp.

We were liberated by the Russian division that arrived and occupied our area almost bloodlessly. It was the 44th Guards Rifle *Baranovitsheiskoy* Division, commanded by Major General V A Borisov, under the overall command of Marshall Rokossovsky's Northern Front Army. This unit had retreated after being overrun in 1941 during Operation Barbarossa, had been reformed, then became a part of the Mongol-horde we later saw with its large component of non-Russian steppes people..

One of the Russian troops, a mortar sergeant then, Vasily Besugly, later retired as an airborne major. In 2001 he came to to several of our meetings at the camp area in Barth with his English-speaking grandson, Andrew Novikov. He came again to Germany in later years to meet with our group of former prisoners. His home is at Ryazan, south-east of Moscow.

At our first meeting at Barth in 2001, each speaker had to be followed by two interpreters to cover English, German and Russian. Quite lengthy, but very interesting to hear all sides. I wish I could have understood everything.

Being outside of the camp I saw much of the Russian army's arrival. They came in like Genghis Khan with all sorts of vehicles, weapons, and huge numbers of dark-skinned, Mongol or Tartar-appearing, steppes-people in leather tank outfits and ragged cloth uniforms for the infantry. They were under the command of the *brown suits,* or Great Russians, from cities like Moscow. I still remember the very loud and continuous noise and the huge clouds of dust.

If Tamerlaine, the Asian conqueror from the Middle Ages, had ridden up on a horse it wouldn't have surprised me. Everything imaginable thundered into town. Some hard to believe in all shapes, sizes, clothes and appearance The nearest thing I can think of was the dust storms of North Africa, but this was much louder.

As I mentioned in the Camp Security chapter, there was more to the *Henry the Butcher* saga for this chapter about the liberation. Henry, last name now long forgotten, an army NCO, had answered Hitler's call for German families to return home to his "Thousand- year Reich". Henry did so, thinking it would improve his life and prosperity as Germany grew as a world power.

Now as the war went rapidly "off the cliff" and downhill, he saw he'd jumped the wrong way. He now, in his anger and despair, picked on everyone he could. He was only restrained by the more senior *Luftwaffe* guards who didn't want to make waves this close to the end of a losing war. As I mentioned previously a young co-pilot in our compound had lived on the same block in Newark and knew Henry well. In my case, I saw Henry and knew who he was, but never spoke to him.

Henry got his punishment, if somewhat more than he'd earned, when he got an arm blown off by a Russian grenade thrown by one of the liberators. I was told he barely survived. A Polish-speaking former NCO prisoner, who could understand the similar Slavic Russian language, pointed out Henry to a new-found Russian friend and told about his misdeeds. I'm sure the American did not expect the result. The Russian casually pulled the pin from a grenade and lobbed it at the civilian-clad former guard NCO. I'm told he almost bled to death before getting quick medical help.

One very strange sight that heightened the Siberian image of the mongol horde, and gave us a huge laugh, was a big Russian "Mama" driving a very large, open, double-wheeled Daimler car with three small kids in the back seat and with all sorts of housekeeping items and a large supply of loot piled and roped on everywhere. She looked tougher than the troops. She was probably following her Russian officer husband, and probably drove across Europe that way following his division. I guess the kids had Slavic "home schooling". She couldn't have been a "camp follower", she would have been the leader.

When a chemical-warfare platoon led by a Russian lieutenant arrived, he quartered his troops in the next farm barn down the road, while he and his sergeant moved into the farmhouse with that family. As soon as he saw me and the American flag on my shoulder, the Russian lieutenant came to shake hands then hug me. He was a brown-suit Muscovite. We became immediate friends with a few words in every language and many "signs". He told me that their actual job was clearing mines.

Several days later he took me with him to attend a dinner with his regimental commander and staff. I was introduced to and shook hands with all the Russian officers present. I don't remember much more, since I understood not a word. I'm sure the others didn't remember much either, because, except for the buffet of cold meat, bread and vegetables, the main menu item was vodka - water glasses of it that you were expected to drink straight down. Very soon we were all sound asleep. I awoke in a huge bed and staggered back to my own farm.

The Russians soon rounded up what probably had been dairy cattle from the various farms, one from each, and herded them to the *Stalag* like a western cattle drive for our use. One of our roommates, Jim Ellis, who'd been raised on a Texas ranch, was the head butcher. The cows were shot by either rifle or pistol, skinned, gutted, hacked into edible pieces, and quickly cooked. We ate the meat as if it were prime rib, even though it was probably from milk cows. Since we hadn't eaten like that for a long time, it was eat, throw up, and eat again until we got over it. Being young, we recuperated quickly and kept eating. We probably didn't gain any weight, because we didn't keep the food down long enough.

Generally the Pow's were ordered to remain at the camp until properly evacuated, but quite a few took off on foot to the west to link up with the advancing Canadian airborne, about thirty miles west, near Rostock. The fellow I'd flown in to Germany with, Lt Pete Abell was one of them. They made it back to Paris long before we did by hitch-hiking first on Canadian airborne, then American, vehicles.

As far as I saw, up until we were finally evacuated, the people were treated normally. As a very sad note, on the peninsula near our camp I came across a young mother's body near the bodies of two young children in a baby carriage. She'd shot them and herself in fear of the coming Russians. We informed the German police who were now under our control and who were also trying to help out. They took the bodies away. The police told us that many women committed suicide.

Stalin wanted our American and British RAF Pow's to be taken back across east Germany, Poland, and Russia by truck or train and to be returned to the US by boat from Odessa on the Black Sea. Our camp leaders knew that Stalin really wanted us taken to Russia as his hostages, so he could extort more advantages and control over Europe from "greenhorn" president, Truman, who still called Stalin "Good Old Joe" even after his butchery of so many millions of his own citizens.

I'm sure Stalin rightly saw Truman as the ignoramus who foolishly gave away so much at the Teheran Conference and brought about the encirclement of Berlin by stopping Eisenhower at Torgau on the Elbe River- with the subsequent need for the Berlin Airlift.

The first American planes that flew into the Barth airfield were communications people, followed by C-46 Curtiss Commando's with medics to take care of the sick and wounded. The nurses brought extra uniforms to dress three German ladies as nurses. These women who had worked in the *stalag's* headquarters had made an agreement with our senior officers to keep our Senior Allied Officer, Col Hub Zemke and his intelligence chief, Col McCollom, up-to-date on all messages and orders that came in from Berlin. In return, our staff promised them safe conduct, and now at the end the word was kept. It also did the same for certain German NCO's and officers who flew out in American uniforms to be handed over to our authorities in France. The Russians were never the wiser, even with all their ID card bunkum.

After a lengthy ten days delay and much haggling at high level the impasse was bypassed and the battle of wills was finally resolved. The Russians required that each of us have an ID card with a signature and a thumb-print. To please them, all of the thousands of cards filled-out had the very same thumb-print, from the same thumb. I'm told all were signed by a small group of people. We're very helpful - ten signers, I heard.

We were flown out of Germany on May 13, 1945 by B-17 bombers from several bomber groups of the 8th Air Forces' First Air Division. There were thirty-five ex-Pow's per plane, but I still got to the glass nose first to see the view along the way. It proved to be well worth the effort.

As the first large incoming formation of tightly grouped B-17's arrived at very low altitude over the camp, about 500 flare guns "liberated" from the German airfield and brought to North One compound for this use, were fired in unison by a flag-wave signal, saying, "Clear To Land". Afterwards, in jubilation we staged impromptu "artillery" battles by firing flare shells back and forth over neighboring barracks. We had large chests of all sorts of shells for the guns brought from the airfield on our large wagon and hauled to our compound. Some were not colored, like roman candles, but were explosive like fire-crackers, so they went off with a big bang. We finally had our own flak.

As the war's end neared many of us talked about burning the entire camp to the ground as our last gesture before leaving. We young "bucks" envisioned our dancing around the flames of this mis-placed "Viking's Funeral" as the Germans do around their giant campfires at Midsummer Night on the longest day of Summer.

We made no secret about the plan during our many discussions, and the word spread widely. However, when the time actually came, at the very end, when we learned about the forced-labor camps and the concentration camps around us, "Hub" Zemke came to our barracks to have a word. He told us that the laborers lived in miserable hovels and if we destroyed the camp these people could not move up to the decent barracks we'd lived in. The colonel told us we could do as we wished without complaint and he'd understand. In the end we agreed with him without disappointment and left the camp intact for those who had not much to go back to.

When we left Barth we were flown, singly, on a low altitude tour of Germany to see the damage inflicted, during our absence, to Berlin and several other large cities. The devastation was so unimaginable that I can still see it in my mind's eye. It was miles of empty honey-comb walls, in burned-out, cities. Unbelievably in the large cities and towns were those burned out by Bomber Harris's RAF, but the small towns seemed totally untouched. We also saw the still visible, and well-kept WWI trenches in NE France, the *dragons-teeth* anti-tank barriers at the Siegfried line, and finally flew on to an Army Air Forces A-26 twin-engine bomber base at Laon, France. One of my good friends, former captain, Maurice Boultinghouse, of Evansville, Indiana, from the 381st Bomb Group, was one of the pilots of the B-17's that came to Germany to fly us back to France. We were taken next by truck to Rheims for a huge welcoming luncheon hosted by SHAEF General Eisenhower, the Allied commander.

After his welcome back speech, and a big meal, we returned to our trucks for the long ride to RAMP (Recovered Allied Military Prisoners) Camp, *Lucky Strike,* on the English Channel coast near the small French town of Fecamp east of Le Harve. The camp was located on a former German airfield on the cliffs looking out over the Channel. There we were debriefed and assigned tents in which to live until we could be given space on a ship returning to the US. We visited all the nearby huge bunkers that the Germans had futilely used while trying to turn back the Invasion, but all their cannons and other equipment had been removed. Now all they did was block the cold wind coming from the sea. We spent several waiting days roaming along the Channel visiting small towns.

The Royal Air Force prisoners from our West Compound at the Stalag were flown directly from Germany to English airfields. Many of the RAF "chaps" had been in German camps for years on end. One friend, Hugh Drummond, who later became the Earl of Perth (Scotland's highest title), had been shot down on a very early mission of the Halifax bomber. He had been a prisoner for four and a half years. Many men were home the same or next day and rightfully. We weren't so fortunate.

As the delays lengthened in our being assigned to a ship to the US, a friend and I, while drinking beer at night in a nearby French inn, decided we couldn't go home without seeing Paris. We walked out to the nearby highway and flagged down a passing DUKW "Duck" amphibious vehicle heading westbound. These are amphibious vehicles used to land troops or haul cargo from offshore vessels. When we showed the troops aboard our German *dog-tags* and told them who we were, we were welcomed aboard by the GI crew, and told we were headed to Le Harve.

We arrived just after sun up and had our new-found friends drop us off at a nearby British army camp. There again our metal "passports" made us welcome. We were given breakfast, a shower and shave, then told they had American OD two-piece uniforms and insignia to outfit us or other walk-in former prisoners. They even had the correct sizes. The good news continued as we were taken to their paymaster, who got out a chart that specified how much cash we could draw and sign for. All this without any official identification. I didn't even have American dog-tags, only the large rectangular German tag with my "2993" and "*Stalag-Luft* 1". We each received the equivalent of $175 in French francs, a British ID card signed by the camp commander, and a jeep ride down to the railroad station. There we found that the British Army had taken over control of the French rail system. The Rail Transport lieutenant in charge of the station gave us a first class compartment for the ride to Paris. Napoleon, we are here!

Now, we were finally and completely liberated; money in our pockets, and headed for Paris and the Big Time. Let the party begin.

When we finally returned to the repatriation camp a week later, poorer, but no wiser, we were told that we'd missed several ships. Makes me think of that WWI song, *How are you going to keep them down on the farm after they've seen Paree,* The Armed Forces Radio station even began broadcasting that there now were "no-waiting" ships available.

Our ride across the Atlantic to Boston was on a ship meant to be, but un-needed as a hospital ship, the *Admiral Mower.* We were *hot-bunked* three high, meaning we had our bunk only for twelve hours, then it belonged to someone else for the next twelve. You had to sleep on that schedule (or on the deck).

We were traveling home with men of the Third Infantry Division who had landed in North Africa, fought across Sicily, Italy, and finally France. I talked to a few who'd made it all the way. Now we finally, had made it "all the way".

15
CHIVALRY

Chivalry is a term that goes back many years and probably existed in fact in one form or another through eons of time before any history was recorded. Dictionaries say it means gallant, courageous and honorable. To most of us the word brings back the tales of the medieval knights whose deeds were supposedly as bright, honorable, and heroic as their gleaming armor, and from the legends in our western world of King Arthur and his men. I'm sure the nations from which many of our ancestors emigrated, in the mostly European world, each had versions of their own. Apparently, however, this form of morality is only accepted in the western and Christian world.

Now, most people, especially those who've never engaged in air combat, find it difficult to believe that such an ethic as this had survived into the modern era and to the wars we fought. Yes, there are also in other parts of the world ethnic and religious divisions that forbid what we recognize as gallantry toward a defeated foe. We've seen much of that in the East. That includes the Middle and Far East, both in the past and in current conflicts. They demand that death, torture, and enslavement be continued as in our more barbaric past. Each of these areas also seem to be dominated by various forms of unquestioning religious fanaticism and by cults that deny chivalry in any form. This is repeated in all of our newspapers and on TV.

Today, during the current wars in the Middle East, this same lack of respect for honor and decency is very evident. Again it seems to be both religious and geographic, as evidenced by the beheadings and torture wallowed in by the various news outlets. These media pretend distaste, with crocodile tears, but eagerly show it for the attention it gets and for the subsequent advertising profits brought in. A growing area of un-chivalrous conduct is also shown in the no-holds-barred fighting, kick-boxing, and almost every sort of mayhem on TV. Perhaps we're moving backwards toward the gladiatorial combat of past ages.

The German Air Force of WWII fought with great skill and unbounded courage, despite many serious handicaps. They fought gallantly, and in most cases with honor, to defend their homeland and nation. They were let down eventually by lack of fuel and a growing loss of experience in the day-fighter force after it was decimated by the ever increasing bomber onslaught.

The night fighter force, on the other hand, was deadly to the British Royal Air Force until the very end of the war, despite the valiant efforts of the RAF Mosquito "intruder" aircraft which hunted German night-fighter aircraft.

After years of large bomber aircraft losses and very high casualty rates, stretching from 1942 into early 1944, American long-range day fighter-escort was belatedly provided for the large bomber formations. Originally, *Area cover,* or the attacks on German fighter bases in France, Belgium and Holland by American P-38, P-47, and later P-51 planes, plus the Spitfire and Hurricane fighter aircraft of the RAF, was the norm. These planes did not have the needed fuel range to make it safely to Germany and back and were a poor pretense at bomber protection. Earlier and better protection of the strategic bomber forces would have allowed these forces to shorten the European phase of WWII by more quickly shutting down the German war industries and national economy - probably with fewer losses on both sides.

Our American air force's top leadership evidently learned little if anything from the early RAF and *Luftwaffe's* catastrophic losses while attempting to conduct daylight bombing without direct fighter assistance. Our Pentagon leadership didn't believe fighter escort was needed. They apparently thought that the heavily armed four-engine bombers could adequately defend themselves. In fact, though, these men were still living in WWI after a new day had dawned, while we were fighting WWII.

During the six-day battle of Operation *Big Week* in February 1944, the German day fighter force lost so many of its best and most experienced fighter pilots and irreplaceable leaders, the die was cast. The decline had already begun. The German day- fighter force began a "death spiral" from which it never recovered, even though it acquired more and better aircraft - in many cases better than the Allied aircraft, such as the BF-262. Although the day fighters did mount many fierce periodic attacks, when fuel supplies permitted, such as the destruction of the 445th Bomb Group in September 1944 at Kassel, these event were on a declining scale. On that day only a few B-24's, out of thirty-some that started out from their airbase at Tibenham, survived to make emergency landings in France and England with much battle damage.

On the other side of the coin, The German air attacks at Guernica during the Spanish Civil War, and the bombing of the so-called "open city" of Rotterdam in the opening days of WWII have been widely condemned as un-chivalrous or barbaric. We know very little about the *Luftwaffe* "volunteers" and the battles they fought in Spain, so I cannot comment on that time period.

Rotterdam is an entirely different story. The Dutch political leaders there, having been given an ultimatum to surrender or face attack, waffled and hemmed-and-hawed until the very last minute before their surrender. The capitulation of the city and the radioed "stand-down" order to the German bomber units, already in-flight, came so late that several bomber groups did not get the recall order in time, or they did not receive it, and continued the attack as briefed.

This caused widespread destruction of property and caused many civilians to be killed, for which the German bomber crews were blamed, but not the Dutch politicians really responsible. There were probably other such instances by the *Luftwaffe*, under orders from their leaders - just as there were for our air force, such as our unnecessary late-war destruction, along with the RAF, of Dresden and Meissen, and other like targets. No doubt, there is a wide-spread difference between political ethics and military honor. The fighting men are caught between the two.

Here we have the lack of honor, or knowledge, by political leaders requiring airmen to destroy part of a country's cultural heritage as revenge. The day and night air raids had no military significance, because there were no military targets involved. "Bomber" Harris of the Royal Air Force Bomber Command is still being castigated, as he should be, but we haven't heard the same about our leader's part, either in these bombings or in their extremely poor judgement about bomber defense.

These things are being mentioned as a preface to the day-to-day reality of our life as prisoners. I was never personally mistreated or bullied. In fact, I was seen by a doctor who examined my damaged ankles and other minor injuries and was talked to by a man I took to be the mayor. Then I was fed dinner before being questioned by the local police chief and by a man in a uniform later identified as the district Gestapo leader. He was a major, or *Sturmbahnführer* of the SS, who gave me a Red Cross form to fill out to let my family know where I was and about my physical condition. At the time, I could not identify his uniform and thought he was a senior police officer.

Sure, we were pushed along when we deliberately dawdled to show resistance, or even misunderstood what was said in German. We traveled in dirty box cars, along with our *Luftwaffe* guards, with little food and water, and with no sanitary facilities other than the tracks when we stopped, but these were the facts of life in wartime. In any military service we call it *exigencies of the service*. Every combat soldier knows exactly what I mean. Our infantrymen in that era had much worse conditions in the field.

Upon arrival at Barth, on 1 March 1944, we marched from the tiny railroad train station about two kilometers out to the camp. This completed a four day "excursion" by small boxcar in each of which there were twelve prisoners and three guards. On arrival we were given decent housing in pre-fabricated military barracks, with bunks in which to sleep, stoves to keep us warm, and reasonable food. It was provided by German issued food and Red Cross food parcels. Food preparation was done in our own mess by competent American volunteers. We later all took turns to volunteer our help, even the Senior Allied Officer, Col Zemke, and before him, Col Byerly. Although we were under the eyes of the tower guards and the security patrols in the compound, our administration was run by our own Allied senior officers who set up normal staff functions. There was minimal outside interference from the *Luftwaffe* prison camp staff.

We were counted at roll call, rain, snow, or shine, in the morning and evening. We had a medical clinic in West Compound that was staffed by British Army doctors and were served by Polish male nurses. More serious medical care was provided by the *Luftwaffe* Orthopedic Hospital at Neubrandenburg, to the south, as long as transportation was available.

In the Spring, or early Summer of 1944, we were given chest x-rays by a self-contained unit mounted in a very large American truck or van that was staffed by a *Luftwaffe* flight surgeon captain and his male technicians. We were told it would visit each of the camps in turn. The US donated the rig with all needed x-ray and processing equipment to Switzerland, which sent it on to Germany for our use. A similar donation of food-hauling 18-wheeler trucks was also later donated in this way, when food could no longer be delivered by train. as I've mentioned in the Cuisine chapter. Allowing these various items to safely get to us was an under-the-table form of chivalry.

I received round-trip mail, plus two parcels, that took almost eight months to travel from the US. Two were from my father, and two were from the wife of the German-American superintendent of the large machine shop facility where I'd been an apprentice when younger. All mail and the parcels arrived safely, if slowly.

These were normal times, well before our bombing campaigns so badly wrecked the German transportation system, that shortages of every kind plagued our lives. We also knew that each shortage meant the end was nearer. RAF prisoner experience and American ingenuity kept us busy if not happy.

I'd also like to mention here that the International Red Cross officials, in their red C-47 *Gooneybird* with the large white crosses on either side, periodically flew in across the Baltic Sea from their base in Sweden, to inspect our camp, before flying on to Berlin. We were fortunate to be one of the few camps that could be reached for inspection on a regular basis. They were our guardian angels in many respects and did whatever they could to intercede when necessary with our captors on our behalf. I have no idea of any complaints made by our senior Allied staff and passed on to these officials, or what response there might have been. We did not receive from the Red Cross any reports about their findings relative to this camp. If there were any, I never heard of them.

As time went by we received all sorts of donations from neutral countries, such as student unions, various governments, etc., and developed a good library and set up schools with all sorts of books and supplies. We organized choirs, a band, an orchestra, a lecture group, put on plays, and began various athletic leagues. Hobbies such as model and tool making were led by experienced volunteers. The results were very evident during the various exhibitions. Most were even visited by German NCO's and officers. There was softball, volleyball, soccer, ice-hockey in late 1944, and hiking while on parole outings. We even had swimming in the Boddensee while on parole outings in the Summer of 1944.

The German Air Force also acquired four American "Andy Hardy" movies in Switzerland to show over the Christmas holidays in 1944. The movies were shown on bed sheets in our mess and in other areas in the other compounds while the whole camp was placed on parole for those days.

There were many other similar things too numerous to count or even to remember. No, it wasn't idyllic, no prison camps are, but our "keepers" were honorable and professional and allowed us as much freedom as was prudent and allowed by their government. The German staff had to walk a fine line with Berlin. Any sign of fraternization by the German officers or NCOs could lead to a courts martial. Most punishments, and these were posted on our bulletin boards, were banishment to the Eastern Front to take part in a losing campaign along with the constant threat of death in battle or slave-labor in Siberia.

The LW camps, at least in our case, were a safe haven for captured Russian flyers, though their officers had to do manual labor as *privates*, since Stalin had refused to sign the Geneva Convention of 1929, which would have given them officer status. I knew several, but could not openly communicate with them when they came to our compound with LW- NCO's to do carpentry and other labor jobs. These men carried the tools, lumber, and whatever was needed by German workers.

We did smuggle these men cigarettes, chocolate, and soap while the guards knowingly and conveniently turned their backs so they wouldn't see the forbidden exchange. These Russians survived if they could get into air force hands. This was also true for the many Polish and other European airmen from occupied countries who flew with the British Royal Air Force, and survived to reach our camp. Most Slavic people, on the other hand, when captured were sent to the slave labor and death camps of the SS if not killed immediately. We were then in the dark about these things.

Although we didn't know about it until after the liberation, there were five slave labor camps, known as Out-camps from the main (KZ) concentration camp at Ravenbruck. They were under SS control in an arc south of us that made parts for Heinkel aircraft, especially for the HE-162 *Volksjaeger* (People's Fighter) jet fighter aircraft. It was one of Hitler's V-weapons, or *Vergeltungswaffen* (Retribution or revenge weapons). Fortunately for our bomber crews the war ended before this plane could make it to combat in force as an interceptor aircraft. The major assembly point was to have been located at the Barth airfield near us. We also knew nothing of this until the war ended, since we'd never seen any flying from the airfield.

The workers, mostly foreign, were beaten, tortured, or killed for any mistakes made that wasted valuable raw materials during hand labor, while not far away, we were treated honorably. This all seems to stem from the German government's vast differentiation between military and political prisoners. I seriously doubt if this same, strange, distinction was made by ordinary German people.

These things are mentioned to show that our German air force captors, in virtually all cases, went beyond what was required or expected to protect us in what was otherwise a very murderous climate. Individual chivalric stories abound, particularly among fighter pilots whose stories I personally heard.

Bob Hoover, then a close friend who lived across the hall from me in room 11, barracks 10, now retired from one of the world's top aerobatic careers, was shot down over the Mediterranean Sea, not far from the German-occupied island of Elba. He was flying a reverse Lease-Lend British Spitfire aircraft while escorting a B-25 bomber formation from Corsica on a bombing raid over Italy. Our P-40 Warhawks were no match for the German aircraft, so the Yanks were given Spits to fly along with all of the needed British flying gear. Being very young and feisty at 20, Bob broke off his assigned escort, against orders, to singlehandedly attack five following Focke-Wulf FW-190 Butcherbird fighters. Of course, they promptly shot him down. He rolled his plane over and fell out of the cockpit as was the standard procedure in the Spit. When his parachute opened, a buckle on one of his British flying boots snagged in the parachute shroud lines and wouldn't pull free. Bob fell the rest of the way to the sea, almost head first, and almost made a hole in the deck of the German air force crash boat coming out to rescue him. Laughingly, they hauled him out with a boat hook and then took him to the German fighter squadron who'd "splashed" him. Instead of turning him in to begin the long trip to prison camp, they kept him for several days of jawing and partying.

To help identify Bob, this is the same Bob Hoover who later was the "chase" pilot for Chuck Yeager at Edwards Air Force Base, though I believe it was then still called Muroc Army Air Base. Later he was also a vice president of Rockwell International.

A similar story was told to us by a young fighter pilot who shared the boxcar with us on the long trip from the *Luftwaffe* interrogation center at Oberusal to *Stalag Luft* 1 at Barth. He was shot down by an BF-109 pilot, then was taken to the squadron that shot him down. He too spent a few days with the German fighter pilots, eating, drinking, and sharing stories. During a Saturday night "bash", he drank little. When the chance came, with a clear head he climbed out of a latrine window and walked in the dark to the flight line. He climbed into a BF-109 fighter, cranked the engine, taxied out, and took off. Everything was fine until the cold engine quit as he cleared the trees at the end of the landing strip. He went down straight ahead into the woods. He wasn't seriously hurt, but the plane was a total loss. It turned out that he'd taken the squadron commander's plane accidentally in the dark. The CO wasn't mad at him, but was unhappy to lose his "beautiful Gustav" - as the "G" model was called. It was the last and best of the Messerschmitt BF-109 series. The combat history of this great fighter aircraft spanned from 1936 to 1945. They were later used by the Israeli Air Force, flying against the Syrians, in taking over their part of Palestine.

One of our roommates, the P-47 Thunderbolt pilot from the 15th Air Force in Italy that I mentioned in the Camp Security chapter, Lt Steiner, had an unusual story to tell us. When he flew his last mission that ended with his capture, he had his crew chief jammed under him in the deep aircraft seat instead of a seat-pack parachute when he crash-landed the plane.. His target was a large headquarters building atop an Italian hill. The two 500 pound bombs he dropped as planned hit the cobblestones in front of the large building and screamed through the large double doors at the building's entrance before exploding. As Steiner pulled up, the plane began to shake apart, hit either by flak or more likely a piece of the building. With no parachute or even time to jump Steiner banked the plane sharply to the left and slid the plane wheels-up into a lower meadow downhill. The German and Italians raced to them almost immediately. Then they began pointing and laughing as first the small fighter pilot climbed out clad in khaki shorts, GI shoes, a gun belt, no shirt, and a baseball cap covered by a radio headset. Then the tall, skinny crew chief who'd been in the seat under him climbed out with a big grin. By putting the seat all the way down, leaving the seat-pack parachute behind, and adding another section of seatbelt, a skinny man and a very small man managed to both jam into the cockpit. How Steiner flew the plane while sitting on the other man's lap I don't know.

With a crew of ten in a bomber we couldn't indulge in such zany flights of fancy as could fighter pilots, because our missions were more structured, longer and had more requirements in destroying enemy assets and capabilities. Now we say that while fighter pilots make movies, we in bombers make history.

Here's another story I recently received from a friend by E-mail: Charles Brown with the 379th Bomber Group from Kimbolton, England, was flying his badly damaged B-17 named *Ye Old Pub* after it was hit by both flak and fighters. With the compass destroyed and clouds covering the sun he was flying eastward into Germany, while he thought he was heading westward toward England.

A young German BF-109 fighter pilot, Franz Steigler, was sent up to shoot down the B-17, but when he got there he couldn't believe his eyes. In his words, he'd never seen a plane in such a bad state. The tail and rear section was severely damaged. The tail gunner was badly wounded. The top turret gunner was killed, with gore across the top of the fuselage. The nose was smashed in and there were holes everywhere. Despite having plenty of ammunition, Steigler flew alongside the plane and looked at Charlie Brown the pilot, who was scared and struggling to control his heavily damaged and blood-stained plane.

Aware that they had no idea where they were going, Franz waved at Charlie to turn 180 degrees left to the west. Franz escorted and guided the stricken plane to and out over the North

Sea towards England. He then saluted Charlie Brown, turned away, and flew back to his base in Germany.

When Franz landed he told his CO he'd shot the bomber down over the sea. He never told the truth to anyone. Charlie Brown and the remnants of his crew told the story at their debriefing, but were told never to talk about it.

More than 40 years later Charlie Brown began looking for the German pilot who'd saved his life and the remainder of his crew. After further years of research he finally found Franz Steigler. Before then he'd never talked about it even at post-war reunions. They later met in the US at a 379th Bomb Group reunion, together with grateful people still alive - because Franz never fired his guns that day.

In the 8th Air Force we'd heard that a badly damaged bomber could drop its landing gear to signal that the crew was going to *bail out*. We were told that the enemy would stop shooting until we were clear. It was just a rumor. But when our badly damaged plane was trailing flame and going down, Pete Abell, the pilot, dropped the landing gear. It was probably one of the few things still working, with the interphone and even the alarm bell out of commission. Opening the nose wheel doors also created an emergency escape hatch for the navigator and bombardier in the nose section.

My only bail out signal was an arm wave from the pilot through his shattered window telling me to come out of the aircraft's nose section. I saw it through where the glass astrodome had been and through the missing pilot's left windshield. They both had been shot away by an BF-109 in a head-on attack just before.

Evidently the fighters honored the "truce". We all made it out safely, except the tail gunner, Lonnie Reaves, who'd been badly wounded earlier and had probably died from loss of blood before then.

Endless stories like this, with every variation imaginable, were our daily fare. After the fright comes the laughter. We were all so young we'd try almost anything, and we had the equipment and the war to do it in. I heard it said that we were so dumb we'd attack a bear with a flyswatter.

Our wars didn't quite turn out as they did in WWI when bombers were slow to learn their trade and when true chivalry flourished between knights of the air. They were said to break off their attacks when one or the other either ran out of ammunition or had a jammed machine gun. As a boy I grew up reading about the famous aces, and I built models of the planes they flew. My first model was of the *Albatross V* flown by von Richthofen, which he flew before he got the red tri-planed DR-1 (Dreidecker) later in the war for which he's so well remembered.

Fortunately, also, we didn't experience the hatred that we've heard about in face-to-face ground combat. There, many of the captured never made it to a prison camp, either on our side or the enemy's. That happened in Europe as well as the Pacific.

When it was possible, I believe that our German air force captors went well beyond the requirements of the rules of war which are unwritten, but well-known. As far as possible, under the demands of their cruel and barbaric government, they treated us with honor and respect while keeping us safe. This was the true shield of Chivalry.

There were, however, exceptions to every rule and large variations among men. On a recent trip back to Germany and Poland we visited the area near Koszalin by the Baltic Sea. It was near here that an NCO camp, *Stalag Luft* 4, was located.

Now we began to hear a far different story. It was one that we'd heard almost nothing about previously. The two former American NCO's, with us, on the trip with their wives, told us a much grimmer tale. On arrival when they climbed down from their boxcars at the railroad station near the *Stalag*, they were forced to run in their weakened state, having minor wounds and whatever, with a bayonet at their backs from the train station uphill to the camp. During their days at the camp, the German *kommandant* mistreated and harassed them personally and through his guards. Food was scarce and very poor.

Along with all their many privations, the prisoners were next forced to march from northwestern Poland to another camp, southwest of Berlin, at Moosburg. This was in the dead of Winter. Many of the prisoners crippled by frostbite and suffering from dysentery and starvation, had died along the many, long, frozen miles. Some of the more fortunate ones were sent on to our camp. Their last stop was at our North 3 compound which at this late date, and with the growing food shortages, had little to offer, other than warmth and shelter, but was vastly better than the conditions they'd recently lived through.

Hitler had ordered that no prisoners were to be released. If camps were about to be overrun by the Russian army, the prisoners were to be shot - not freed or abandoned. This order was also posted on our bulletin boards at *Stalag Luft* 1. At first there was much consternation, but it gradually died away when it became apparent that due to our location there were Allied armies closing in from both the east and west. There was also the determination of our Senior Allied Officer, Col Hub Zemke, that we wouldn't leave our camp area regardless of any order from Berlin. The details are in his book, *Zemke's Stalag*.

In all of the prison camps east of us from East Prussia and Poland, with the unstoppable approach of the Russian Army, it was ordered that the prisoners be made to walk back to Germany under guard. It was a 'death march' every bit as bad as the Bataan Death March, even though now it's largely unpublicized and almost entirely unknown in this country. Here the forced march was in the cold of winter instead of tropical heat as in the Philippines. We're told that the number of deaths was at least as high.

One group walked for 85 days without washing, shaving, changing their clothes, or brushing their teeth, through freezing rain, ice and snow, and then thawing mud. They lived the entire time in the same dirty clothes and wore everything they had. Even for those who had a shorter distance to walk, and fewer bad days to live through, the conditions were terrible. Without shelter the men and guards slept in barns, culverts, even in open fields, huddled for whatever warmth and shelter they could find. Food was scarce and hunger was continuous. A book, *The Last Escape,* by John Nichol and Tony Rennel, copyright 2002 by Penguin Press, which I now have, is the only book I know of that covers the suffering and deaths of this terrible time for Allied prisoners of war. Americans need to be aware of this.

Stalag Luft I was sheltered by geography and the cold-blooded courage of Colonel Hubert L Zemke former CO of the 56th P-47 Fighter Group who was our Senior Allied Officer. We had the Canadian airborne troops driving eastward towards us from Rostock, the Russian 44th Guards Rifle Division driving westward from Stralsund in the east, and various Russian tank divisions circling around the north side of Berlin south of us. On the north we had our backs to the Baltic Sea. We could not have been moved in any direction. We were more fortunate than we knew.

Although our acquaintanceship with the German air force NCO's and officers was necessarily limited by restrictions based upon our security needs and their need to avoid any semblance of fraternization, we both learned a joint respect. Their thinking, for the most part, was similar to ours in honor and in a shared sense of humor. We survived, if not prospered, under the shield of Chivalry.

To a certain extent, I think we see what we expect. As a slightly longer-termed American prisoner in Germany, who had earlier seen times of better food, living conditions, and more cooperative German captors, I knew that many if not most of the changes for the worst were more like the symptoms of Germany's collapse and the specter of defeat, than of a lack of chivalry.

I completely sympathize with those fellow kriegies who were forced into the long, terrible, evacuation marches by Hitler's order from camps in the east about to be overrun by Russian forces. I also understand the feelings of POW evacuees who arrived at our camp in the later days of the war. They had spent far less time in captivity than some of us did, but were captured late in the war and never saw the "normal" times and activities we enjoyed earlier during the war.

These men felt now that they were being deliberately mistreated when they arrived and had to share our much poorer living standards.

Unfortunately, some of the German air force men who had guarded us honorably, or even chivalrously, were sent at war's end to labor and die in Russia's own slave labor camps, the *gulag*.

16
AIR EVACUATION OF STALAG LUFT 1

8TH AIR FORCE PRISONER OF WAR RELATES EXPERIENCE
DONALD M MENARD, 398BOMB GROUP (H), 603RD SQUADRON
4811 ANGLETON COURT, BATON ROUGE, LA 70817

In response to Newt Moy's letter to Mail Call in the June 00 issue in which he mentioned participation in the air evacuation of *Stalag Luft* 1 at Barth on the Baltic coast of northern Germany. What I relate here is by no means all events and details of what took place during the two weeks from May 1, 1945 through the final lift off on May 14th. Most of what I submit comes from notes, memory and some excerpts from Col. Hubert Zemke's records and his writings. Colonel Zemke, Commander of the 56th Fighter Group, was the Stalag's Senior Allied Officer and as such was the commander of Provisional Group X. This group represented all of the Allied prisoners before the Luftwaffe staff operating Stalag Luft 1.

Due to the proximity of the Russian forces to us in late April, the German staff asked for a meeting with Colonel Zemke. The purpose of this meeting was to propose to leave Stalag Luft 1 to us and they would head west to contact British forces who were within 70-80 km of Barth. This was agreed upon. At approximately midnight of May 1, the Germans' departure took place. The camp population was not aware of this operation until daylight when the Stars & Stripes and the Union Jack had replaced the swastika. The flags had been sewn with scraps and smuggled for quite some time.

Since the facility was not marked and it was unknown whether the Russians were aware of our existence, a scouting team of our now liberated prisoners who spoke German and Russian were sent toward the east to intercept forward elements of the Russian army. From our location we could easily hear what was probably artillery fire. Hence the importance of making contact. The original party arrived late on May 1st with a Russian scout who was taken to the Stalag. His information was that the main force of the Russian 65th Army was near Stralsund, about 20 km east of Barth. The ranking Russian officer for this group was a Colonel Zhovaniak. Most interaction between ourselves and the Russians was between Col. Zemke and Col, Zhovaniak, later joined by his superior, General Borisov.

Much to the chagrin of almost 9000 liberated POW's, it took at least two weeks to clear our repatriation by air. The Russians' idea was to send us by land to Odessa on the Black Sea, then by ship to the United States and the United Kingdom. This was rejected and further negotiations followed.

Finally on May 11th came authority to use the airport adjacent to our camp for evacuation on the 12th and 13th, within specified hours. A 20 km corridor for travel over Russian occupied territory was designated. By then everyone had their meager belongings together. The boarding plan was to first take hospital cases and the sick. Next were the British who for the most part had been incarcerated longer than any of the prisoners, back to 1940. Prisoners were marched in barracks groups on the trip to the airport to avoid cluttering the roadway and loading area at the airport.

In the early afternoon of the 12th the first B-17 arrived. Aboard was General William Gross,

Commander of the 1st Air Division of the 8th Air Force. A C-46 followed with officers from General Eisenhower's staff at SHAEF headquarters. Two more B-17's arrived with communications equipment to establish links with 8th Air Force. Later in the afternoon 30-40 more B-17's arrived. The fuselages of these aircraft were decked with what appeared to be wooden packing crate material or something similar to plywood. This formed somewhat of a level floor in the fuselage thereby accommodating 25-30 POW's including the few packed in the radio room.

Early on the morning of the 13th the sick and wounded were evacuated in six C-46's. Dozens of C-46's and C-47's joined the B-17's in this process. The aircraft crept very slowly while the men jumped in and piled up in most cases. Being from the 398th Bomb Group, 603rd Squadron it was a pleasant surprise for me to board in the afternoon in an aircraft not only from my group but my squadron as well. I have no idea how many groups participated. I do remember seeing some from our 1st Air Wing, the 91st from Bassingbourne and the 381st from Ridgewell. I know there were others but I do not know and have not researched it. According to my notes, the total number evacuated for the second day, May 13th, was 6250.

On the morning of the third day, the 14th, the remaining men were all flown out aboard B-17's. The last prisoner to board was Colonel Zemke after having met with General Borisov and Colonel Zhovaniak. According to Colonel Zemke's records, a total of 8487 had been evacuated. The return trip to France did not seem to be through a corridor. The pilot took us over several target cities to view the destruction. Our particular aircraft landed at an airfield at Laon, France. We were transported by truck to what appeared to be a troop carrier base near Rheims. From there we were transported by C-47 or C-46 to RAMP Camp Lucky Strike near the small town of Fecamp in the general area of Leharve. At Camp Lucky Strike we were debriefed, issued clothing and for those of us who had experienced various stomach disorders were given diets like soft food with liquid high calorie drinks. My particular stay there was about eighteen days. My trip back to the states was aboard a very new and swift Navy troop transport. The trip to New York was 4 ½ days. From New York we went to New Brunswick, NJ, and Camp Kilmer to pick up travel orders and summer clothing. After that, travel was to Camp Shelby, MS. There we picked up pay vouchers and furlough orders and base assignments after the furlough. From there it was on to my parent's home in Lafayette, LA, arriving June 15th.

From the December 2000, Vol. 00, Number 4, 8th AF News.

17
Barth Airfield

The small, one short paved runway of today's commercial Barth airfield began its existence in the 1930's as a Luftwaffe airbase, called Fliegerhorst Barth. Originally it consisted of a large, well-drained, grass-covered field with concrete paving provided only for the ramp areas in front of the hangars, and at the refueling points. This was common to all countries during the birth and early growth of aviation, including for both both military and civilian aircraft uses. Later during the approach to WWII a single runway was built to accommodate the ever-growing weight and larger sizes of the newer planes. I'm sure weather with its seasonal mud would also have been another determining factor.

All of the original airfield buildings to the north of the current runway were torn down and destroyed by the Russian occupiers who left only scattered debris and a few foundations on the previous northern side. Now there are only small buildings on the opposite, or south, side of the runway for a commercial aviation terminal, although on our last visit in 2006 we saw the beginning of construction for a new air terminal, other building improvements, and various expansions underway.

On 13 June 1938, soon after the airfield was completed, Hitler, the leaders of the German Air Force, Army, Navy, and their many ministry officials visited the airfield to observe pre-war air-power demonstrations with the local air training officials.

Air power was recognized in Germany very early as the keystone of military might, on a par with Army and Naval power. Although, because of the large cost and long lead time for construction, sea-power was gradually limited to focus upon a large offensive fleet of submarines. The lower cost and shorter time to combat readiness meant that only a limited surface armada would be possible until other more pressing needs were met. The few surface battleships built were smaller, faster, and with heavier fire-power, but with only light armor protection, such as the *Bismarck*. Its "Achilles heel," was shown by the torpedo-jammed double-rudder system which allowed the much slower British Navy to catch and sink it. For a nation with a relatively small population, Germany created a larger and more modern air force and army more quickly than was realized, even by the US and Britain.

Charles Lindbergh, still a hero from his earlier Atlantic crossing flight, was proudly invited by *Luftwaffe* commander, Hermann Goering, to visit Germany in the late 1930's and view the length and breadth of his emerging large air fleet. Lindy accepted, saw what had been built, and then appeared before Congress to pass on what he'd learned with an eye toward encouraging the US to modernize and enlarge our woefully lacking Army Air Corp's equipment, size and training. Even though Lindbergh was a colonel in the reserve, a peeved Roosevelt would not let him be called to active-duty for service in later WWII. Instead, Lindy spent the war as a technical representative for an aircraft company furnishing the F4-U Corsair planes for the Marines and Navy.

As the years passed rapidly in the ever faster build-up to WWII, the Barth aerodrome took on many more duties and far greater responsibilities. Hangars, barracks, and all of the other needed buildings and facilities were erected and constructed to house the growing number of personnel and aircraft to provide for their needs and coming war mission. Then it was probably as modern as those in far bigger towns. Now after destruction during the forty-year occupation by Russia, it

HEINKEL HE-162

VOLKSJAEGER

PEOPLE'S FIGHTER

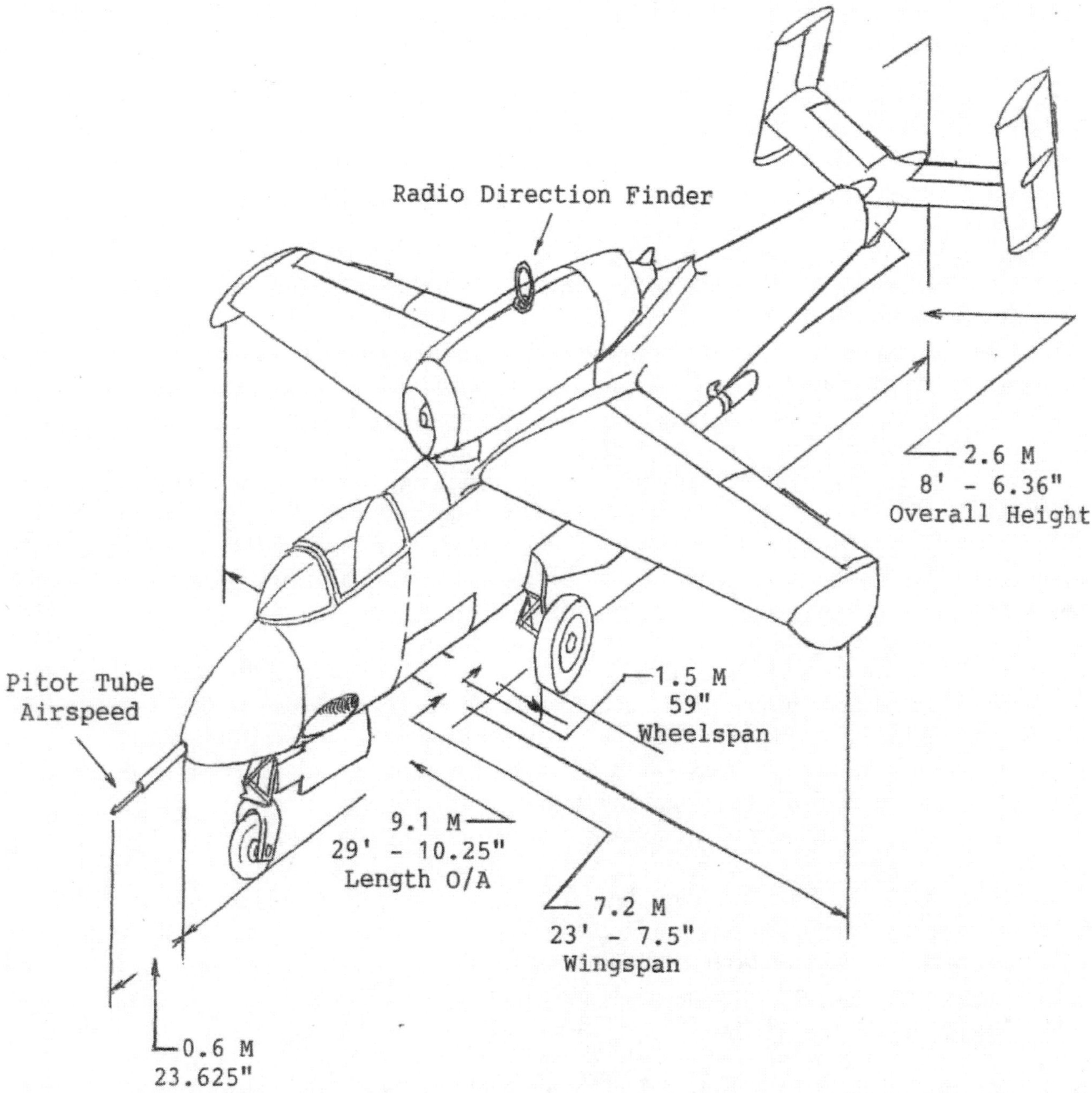

Archiv Albrecht/BERLIN
Dr. Martin Albrecht, Phd

Volksjaeger Drawing from
Archiv Albrecht/ Berlin

is awakening from its many decades-long sleep - this time as a commercial air hub for the large surrounding area.

During the airfield's span of years as a military airfield, many types of flight training were conducted for, first, instructor groups and then transport crews that formed the groups flying the Junkers JU-52 tri-motored transport plane, which they later lovingly-named *Iron Annie*. Aircraft that were then considered heavy bombers such as the Heinkel HE-III made an early debut, followed by heavier twin-engine fighters such as the ones we called the Messerschmitt ME-110, (but whose actual name was BF-110, since the BF really stands for Bayerische Flugzeugewerke, or Bavarian Aircraft Works). Various models of dive-bombers; and the small single-engine transports, or executive aircraft such as the BF-108 followed. Originally the BF-108, the *Taifun* (Typhoon) was a tail-wheeled aircraft, but it was later changed over to a nose landing-gear craft that became the BF-208.

The original BF-108 was the predecessor of what later became the BF-109 series of fighter aircraft that began its duty in the Spanish Civil War and soldiered-on, excellently, through WWII. Though few American know of it, this old eagle, along with its valiant partner, the Focke-Wulf FW-190 Butcherbird, made its fighting days last through the wars that founded Israel

A non-flying, but restored for display, BF-208 aircraft built under license by Renault near Paris is on display at the Warbirds Museum on TICO Airport at Titusville Airport near the Space Center in Florida.

Unfortunately for Germany, Hitler's "tunnel-vision" was able to see only fighters and dive bombers. Most were used at one time or another in both combat roles. Even though the Heinkel HE - 111 was normally used as a level bomber, under Hitler's demand it too had to be capable of dive-bombing. What was an even more ridiculous part of this policy was to require that the Focke-Wulf, FW-200, Condor, that was actually a four-engine transport or air-liner, to be capable of dive-bombing. No real long range strategic bomber that could bomb the tank and artillery factories beyond the Ural Mountains was contemplated or built. This one-eyed approach to air power led Germany to certain destruction - as it did the Japanese in their one-shot, win-or-sink war in the Pacific.

Does that remind you of our General "Hap" Arnold, who thought our bombers didn't need protection against enemy fighters - and who didn't get escort fighters to us until the later part of the air war in Europe - until after many thousands of bomber crewmen were killed, captured, or wounded? When you hear the fighter pilot stories today you never hear how belated their help was.

The war years brought a whole range of aircraft to Barth, almost a catalog of German military planes of that period. Many of these operational planes were used as fighter-bombers (Jagdbombers, or ' Jabo's) in an attempt to repel the Russian ground forces over-running north German lands and our camp area from the East. These planes included BF-109's, FW-190's, Heinkel 219 Night-fighters, Junkers JU-88 Schnellbombers (Fast bombers), and even an Arado Blitz, a light twin-engine jet bomber that was then far ahead of its time. We saw only one in flight, so it might have been a visitor.

As mentioned in other chapters, from our North One compound we could hear the various aircraft engine sounds and sometimes see the tops of the aircraft vertical stabilizers as they taxied to and from their runway. We counted them off on their missions, then counted them back-in during their landing patterns on return, to see how many had been lost in battle with the Russians. Because we heckled the men in the guard towers about the losses and during the frequent air raids, we were later ordered to stay in our barracks during theses events, or be shot. Two men in our compound were shot that way. One man died, as mentioned in another segment.

During this *Gotterdammerung*, or twilight-of-the-gods, as the war's end neared, there were even young flying cadets being given advanced combat crew training in the Focke-Wulf, FW-190 single engine fighter aircraft and other such fighters at the base. Fuel and experience became very limited and training was short before these young men were pushed from the nest and into combat action before their feathers grew out.

For air defense the airfield had rapid-firing, light anti-aircraft cannons such as the *Vierling*, a four-barrel 20mm weapon, and 37mm single and multiple barrel cannons of the Bofors type. Light and heavy machine guns rounded-out and filled the gaps. I did not hear the "bark" of the famous 88mm flak cannons that were used widely for air defense - and I did not see any of their units at the airfield area when we marched from our camp to the field to climb aboard the B-17's that ferried us to Laon, France, at war's end.

On one afternoon during our stay at the camp we saw a P-51 Mustang shot down while making a target-of-opportunity attack after being released at the end of his bomber escort duties. We could hears his guns firing as he dove down while attacking parked aircraft or other ground targets. As he pulled up too soon after the firing pass his aircraft was hit by this flak. A white cloud of vaporized ethylene glycol coolant from the Rolls-Royce *Merlin* engine left a large white streamer behind the aircraft as the pilot pulled-up while trying to gain enough altitude for bailout. Instead of trying to "belly-in" his plane he jumped too close to the ground and was killed on impact before his chute could fully open. He was given a proper military funeral at a local cemetery with a German-made American flag

Although we Allied prisoners had no knowledge of it, a *Konzentrationslager* (KZ), or concentrations camp, was also located on the airfield from November 1943 onward to the end of the war. It was one of a five camp systems of forced-labor KZ's staffed primarily by French and Polish conscripts and operated by the SS as a sub-camp group controlled by the notorious Ravensbruck main KZ. These camps were located on the mainland in an arc south of our camp's peninsula. The workers made, almost by hand, aircraft parts for the Ernst Heinkel Flugzeugwerke GMBH (Ernst Heinkel Aircraft Works Corporation) Rostock Plant, called the Mullerwerk. We were told that the SS tortured and killed many of the starving inmates if they attempted to escape, or if they made mistakes, or slacked-off in their work. There is a monument to memorialize the deaths of these men in a small park on the road that leads south from Barth. We've visited it each time we have toured our old *stalag*. There is also a similar memorial for the Russian laborers who died there.

The staff, workers, and slaves were at the Barther Fleigerhorst attempting to assemble one of Hitler's "last-ditch" weapons, the HE-162 Volksjaeger, or People's Fighter. It was one of Hitler's last "V" or *Vergeltungswaffen* (Retribution or revenge weapons). It was a single-seat, tricycle landing gear, twin-tail, single-engine jet fighter. There wasn't even time to assemble this as-yet unproven last-defense aircraft from the finished parts on hand and get it into the air, much less train pilots to fly it. How it would have managed in combat as an interceptor fighter aircraft against our ever-growing bomber fleets is a mystery with no solution.

The war ended for us when the German *stalag* command turned over the area to senior American and British officers. German air force personnel and their civilian staff left the area well before the Russians arrived with the advance elements of the 44th Guards Rifle Division. In the various KZ's nearby the SS staff and guards melted-into-the-night and deserted freeing the almost skeleton-like prisoners. Townspeople from Barth, doctors, and nurses rushed to aid and to keep alive the remaining inmates. I don't have any of the details, but I'm sure that historian Helga Radau in Barth has an accurate account of their efforts, even though she was a child at the time in 1945.

Since this area was handed over from American control to the Russians upon their arrival, probably all of the aircraft drawing, parts, and partially assembled HE-162 aircraft were transported to Russia to begin their own jet aircraft program. Somewhat later the far left-wing British Premier, Wilson, also gave the Russians the latest and best British jet engine to help speed them along. That culminated in the Soviet MIG-15 fighters which I encountered over North Korea during that war while flying B-29 Superfortresses, on many bombing missions

We've been told that the Barth airfield was to have been the major assembly point for this aircraft.

Most of the details about the airfield, its war missions, and the belated aircraft production attempt are covered in the German-language book, *Flugzeuge aus Barth,* (*Aircraft From Barth*), authored by Dr Martin Albrecht, Phd., Director of the Documentation Center in Berlin and Helga Radau from Barth. Ms Radau from a very young age, even under the Russian occupation, was and is the camp and area historian.

With Dr Albrecht's kind permission I have included sketches and a photo of the Volksjaeger, or People Fighter, The Heinkel - 162.

Heinkel HE - 162 Volksjager

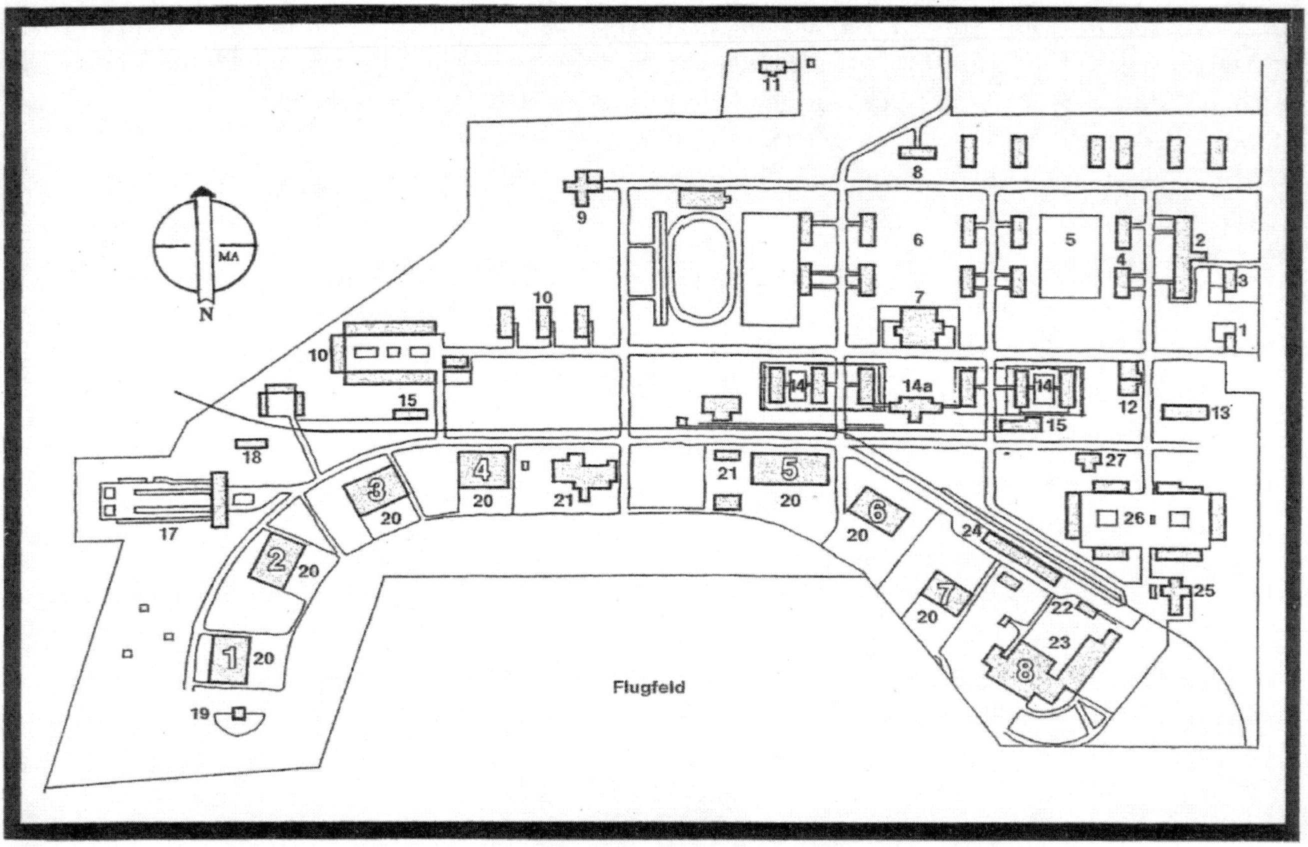

FLIEGERHORST BARTH 1935 UNTIL 1945

Ernst Heinkel Flugzeugwerke Gmbh, *Muller Werk* 1943-1945

1. Main Gate & Guardroom
2. Const HQ, Mail, Barber
3. Inspections Office
4. EM and NCO Barracks
5. Exercise Field
6. EM and NCO Barracks
7. Team or Club Canteen
8. First Aid Clinic
9. Officer's Club and Mess
10. Officer's Quarters
11. Airfield Commander's Home
12. KZ Headquarters

14a. Canteen of the KZ
15. Hot House
16. Parking Garage
17. Machine Gun Test Ranges
18. Chicken House & Pig Pens
19. Acft Engine Test Stands
20. Aircraft Hangars- 7 each
21. Flight Operations and Commander
22. Weapons Storage and Shop
23. Hangar
24. Railroad Station
25. Worker's Canteen
26. Motor Transport Shops

KZ and Muller Werk

1-8, Halls for Production of Aircraft Parts, Aircraft, and Components

12. HQ of Muller Werk and and KZ Commander

14. Quarters for KZ Prisoners In 3 Eastern Barracks for Males & 3 Western for Females

14a. Canteen for KZ Workers

KZ- Abbreviation for Konzentrationslager or Concentration Camp

Archiv Albrecht/Berlin
Dr Martin Albrecht, PhD

18
Afterword

As I noted in the Liberation chapter, Stalin refused to release the POWs at *Stalag Luft* 1 after the war had ended in our North German area. Our flight out on 8th Air Force B-17s to France was delayed for 10 days while he tried to force the "not too bright" Truman and his equally backward cronies to make further territorial and political concessions about the defeated German nation and people.

However, this was not the first or the last effort on Stalin's part to grab greater and more far-reaching control over Europe and its people. Most of these efforts are not known even to this day, because the American and other allied governments have continued to foster the myth that the Russian government is an ally. Nothing could be further from the truth.

In the first major event, before the German armed forces attacked Russian armies across the Polish border in *Unternehmen Barbarossa,* Stalin had 149 divisions on the German border being assembled and outfitted to attack Europe in 1941. This is the primary reason why German forces so easily captured 3.4 million men and vast amounts of military supplies with such minor losses to their own units. All of the details, chapter and verse, are covered in great and accurate detail by Prof. Constantine Pleshakov in his book, *Stalin's Folly*. Having access to the Moscow archives and knowing the language enabled the professor to give us a clear picture of the actual events as they occurred, showing the Russian military units involved, the command structures, and the names of the military leaders.

The second similar event began as World War II ended -- and as Russia began to regain its military strength. Truman, with absolutely no idea of what dangers we faced in a post-war Europe had allowed our then powerful and well trained military forces to disband and leave the continent in the hands of the weak and inexperienced occupation force. He again was the "pigeon." Truman's apparent weakness and vast lack of knowledge led Stalin to believe that Western Europe could swiftly be overrun before we could react.

Stalin with his one track mind began to assemble another great army to march westward, almost unchallenged, across Europe -- this time to the Atlantic Ocean.

Fortunately, there was a successful challenge and blocking force, and a very small but mighty air armada. No, again you've never been told about it by your government. The newly created Air Force created a B-29 air division at MacDill Air Force Base, Tampa, Florida, under Gen. Kenney. It consisted of only the 306th and 307th bomb wings, but these units were sent to England equipped with nuclear weapons. Our government still keeps silent, preferring the myth that the Russians were still (if ever) our allies. This force, later with B-47s and B-52s in the "third strike," gradually grew into the weapon that again stopped Russia in the Cold War. Later the U.S. Air Force with its ICBM rockets and the Navy with its submarine "Boomers" completed the "triad."

No, I wasn't then in the bomber forces, back in 1948 and didn't know the reason for the hurried assembly that saved Europe from disaster and world war three. I was with the 5th rescue squadron -- also at MacDill Air Force Base. Our job was to fly long-range patrols far out into the Atlantic to provide rescue capability for the B-29s that checked in with us as they passed to and from England. However if a B-29 had to "ditch" at sea -- and if we had to land to pick up the downed crewmen with our war- weary Navy "cast off" SA-10 Catalina flying boats, we'd probably all gone to the bottom together. It's a good thing that the Boeing B-29 bombers completed their missions safely, because we really couldn't have provided much help with the poor planes we flew.

Later I flew the same B-29 models with the Strategic Air Command's 98th bomb wing during the Korean war on many combat missions, that helped stop North Korean forces from over-running South Korea. The North, under communism, is still wallowing in poverty.

Yes, there are repetitions throughout the book - they are an attempt to help maintain understandibility and continuity.

Appendix and Photos

Wer erinnert sich?

Am 30. Dezember 1943 fand ein Groß-einflug der 8 th US-AAF auf Frankfurt/Main statt.
Anflugweg: Kanalküste, Nordfrankreich, Luxemburg, Trier.
Die III./JG 2 „Richthofen“ hatte über Nordfrankreich, speziell im Großraum Reims, Luftkämpfe.
Bei diesen Luftkämpfen mit Viermots ist am 30. Dezember 1943 bei NOGENT/REIMS, der

Uffz. Eberhard BORCHERT, 9./JG 2 „Richthofen“
* 26. 8. 1921 in Lötzen / Ostpreußen

gefallen.
Uffz. Borchert war vom 6. 1. 1943 bis 8. 5. 1943 bei der 9./JG 2, vom 9. 5. 43 bis 26. 10. 43 bei der Frontfliegersammelgruppe Quedlinburg und dann wieder vom 27. 10. 1943 bis 30. 12. 1943 bei der 9./JG 2 „Richthofen“.
Wer erinnert sich an Uffz. Borchert? Wer kann etwas über seinen Tod aussagen? Wer kennt die Grabstelle?
Hinweise bitte unter Az.: ES/303 an: GdJ-Sozialreferent, 2322 Lütjenburg, Postfach 72.

Translation of the German Text...

Who remembers him?

On 30 December 1943, a great air-fleet of the USAAF arrived over the Frankfurt-Main area. The fleet approached the Channel Coast, Northern France, Luxemburg and Trier.

The 3rd Squadron of *Jagdgeschwader* 2 (Fighter Wing 2) *Richthofen* had the responsibility over Northern France especially in the Reims area air battles.

At these battles with 4-engine aircraft on 30 December 1943 at Nogent/Reims, the Unteroffizier (Luftwaffe corporal - American, sergeant) Eberhard Borchert of the 9th Squadron, JG-2 *Richthofen*, born 26 August 1921 at Lotzen, East Prussia, fell in battle.

Uffz Borchert served from 6 January 1943 until 8 May 1943 with the 9th Squadron, JG-2; from 5 September 1943 until 26 October 1943 at Front Flyer's Collective Group Quedlinburg; then from 27 October 1943 to 30 December 1943 served with JG-2 *Richthofen*.

Who remembers him, Uffz. Borchert? Who can say something of his death? Who knows the gravesite?

Information, please, under AZ.: ES/303 an GdJ-Socialreferent, 2322 Lütjenburg, Post Office Box 72
No date is given for the publication.

Unteroffizier Eberhard Borchert was the Focke-Wulf pilot who attacked our aircraft on 30 December 1943. After we both fired at each other his plane nosed over and dove down to crash into the English Channel. My nose-turret was shattered and I was wounded in the left leg. The added battle damage knocked out the number four engine causing us to "ditch" about five miles offshore from Beachy Head west of Eastbourne, England.

Another Pilot Missing

Although Unteroffizier Eberhard Borchert seems to have been the most likely victim in our 30 December 1943 clash over the English Channel, a second flyer from Jagdgeschwader 2 went missing over the same general area on the same afternoon. He was Unteroffizier Josef Schirmer pictured here with a BF-109 he flew previously with Jagdgeschwader 54. After he was transferred to Jagdgeschwader 2, "Richthofen," he made the transition to the Focke-Wulf FW - 190 fighter aircraft with which they were equipped. There's no way to know positively which pilot was my opponent.

Historians speak of "the fog of war" to explain the very great difficulty of learning later precisely what happened in any given battle, because even the survivor's memories at the time are scrambled and disjointed. It's a bit like a car wreck and often happens just as quickly. In this case, imagine trying to capture an exact account after the intervening sixty-eight years that have elapsed since this happened. Fortunately German records-keeping and its historians are among the most reliable.

Group photo of German pilots from Jagdgeschwader 2, *Richthofen* (Fighter Wing 2) posing with an FW-190. The pilot on the extreme left is Unteroffizier Eberhard Borchert. (Please see facing page for more on this pilot.)

PLACING THE WREATH AT THE TOMB OF THE UNKNOWN SOLDIER

On 14 May 2011 while traveling with the Volusia Honor Flight from Daytona Beach airport to Washington, DC to enable World War II veterans to view the memorials of World War II and the later wars, we were hosted there by U S Representatives John Mica and Sandy Adams. Our flights to and from the District were provided by the Rotary and other business and patriotic clubs in the Daytona and Deland areas on our own U S Airways aircraft.

This trip and the luncheon on the previous Saturday at Daytona State College were handled extremely well in each and every endeavor. Every detail from start to finish, was handled from early morning to a late night one-day return from the flights, the transportation in Washington, the monument tour, the luncheon at the House of Representatives' Cannon Office Building, and the police escorts for our five busses. We felt very blessed by the overwhelming recognition by people all along our route of travel. When leaving in the early morning from Daytona Beach airport and the return later that evening we were met by hundreds of people not there to catch a flight, but who came there to greet us and shake our hands. They ranged from older people to even young tots.

There were 103 veterans and 52 escorts. My escort was Steve Rider, there to escort his father from Lake Placid, FL and me from Titusville, FL.

When we reached Arlington National Cemetery and waited for the wreath ceremony to begin, after a short chat with our congressional representatives, Representative John Mica told me that 93 year-old former soldier, Marcel Soliefier and I, both also former prisoners, would make the wreath presentation jointly - one on each side of the wreath.

This is an honor normally bestowed only upon military and political leaders, generals and presidents, on very special and solemn occasions. It is an understatement to say that this is the greatest honor of my life - to symbolically, before this large audience, to help honor all those who have given their lives down through the centuries for this great country.

A Vice-Admiral from the Pentagon was my escort as I marched forth in a green rain jacket and completed this unsought but beloved task. It was also a very humbling experience.

Maj. Reus placing the wreath at the
Tomb of the Unknown Soldier,
Arlington National Cemetery,
Washington, D.C.

Top Left:
Maj. and Mrs. Joe Reus at Herbstein. Maj. Reus' plane exploded over this villiage on 24 Feb 1944.

Bottom Photo:
From Left:
Ranier Dietz, owner and chef at his Johannesberg Inn; Joseph Reus; Horst Jeckel, Historian; Shirley Reus, and Robert Keller, Historian.

Top:
Maj. Joseph Reus on left with Helmut Roth, who flew the BF-109 and was later a POW in France. Mr. Roth died January of 2011. The photo was taken at the Memorial at Ludwigsau honoring fallen from both sides.

Bottom Right:
Dr. Martin Albrecht, Director, Berlin Documentation Center (Archives)

Bottom left:
From Left: Gerhard Kott, an FW-190 fighter Ace; Horst Jeckel, Historian; Jim Baynham (from Texas, who flew one of the B-24s shot down during the ill-fated Kassel Mission which almost wiped out our 445th Bomber Group Sep 1944) and Maj. Reus.

Beautiful color shot of a BF-110